EXPERIMENTATION AND UNCERTAINTY ANALYSIS FOR ENGINEERS

EXPERIMENTATION AND UNCERTAINTY ANALYSIS FOR ENGINEERS

Hugh W. Coleman
W. Glenn Steele, Jr.

Mississippi State University

WILEY

A Wiley-Interscience Publication

JOHN WILEY & SONS

New York / Chichester / Brisbane / Toronto / Singapore

Library of Congress Cataloging in Publication Data:

Coleman, Hugh W.
    Experimentation and uncertainty analysis for engineers/Hugh W.
Coleman, W. Glenn Steele, Jr.

       p.   cm.
    "A Wiley-Interscience publication."
    Bibliography: p.

    1. Engineering—Experiments. 2. Uncertainty.   I. Steele, W.
Glenn.   II. Title.
TA153.C66 1989
620'.0072'4—dc19   89-5550
   ISBN 0-471-63517-0   CIP

Printed in the United States of America

10 9 8 7 6 5 4 3 2 1

To my wife, Anne; sons, Matt, Andy and Jeff; and parents,
Marion and Mable Coleman

H.W.C.

To my wife, Cherie; daughters, Amy, Holly and Carrie; and
parents, the late Wilbur and Mary Steele

W.G.S.

CONTENTS

PREFACE

Our objective in this book is to present a logical approach to experimentation through the application of uncertainty analysis in the planning, design, construction, debugging, execution, data analysis, and reporting phases of experimental programs. It is intended that the book will be appropriate for use in an upper-level undergraduate course, in a graduate course when supplemented with readings from the literature, and as a continuing education resource and reference for practicing engineers.

Although the book is written with a mechanical engineering perspective, we have attempted to achieve a generality that will have appeal to a range of engineering disciplines and to the physical sciences. We assume a familiarity with mathematics and science typical of junior/senior undergraduates in engineering and the physical sciences and also an introductory knowledge of the basic principles of transducers, data acquisition, etc.

This book is unique in that the estimation and propagation of both precision (random) errors and bias (fixed) errors are considered as opposed to the previous practice of assuming that bias errors are negligible or have been eliminated by calibration. We discuss the procedures for handling small sample sizes, which require use of the t-distribution, and for handling the important practical cases in which bias errors in different variables are correlated. Our presentation follows and explains the approaches in the ANSI/ASME Standard on *Measurement Uncertainty*, which was approved and adopted by the American National Standards Institute in February 1985 as an American National Standard and was issued by the American Society of Mechanical Engineers in April 1986.

The seven chapters in the book can be divided into two parts. In the first part, Chapters 1–4, we develop the methodology for properly considering the uncertainty in measured variables and the manner in which these uncertainties propagate into the uncertainty in the result of an experimental program. In Chapters 1 and 2, we introduce the definitions, concepts, and statistical ideas necessary to consider the estimation of precision error, bias error, and overall uncertainty in a measured variable. In Chapter 3, the application of general uncertainty analysis in the planning phase of an experimental program is developed and illustrated. Detailed uncertainty analysis, which considers separately the propagation of precision errors and the propagation of bias errors into the experimental result, is developed in Chapter 4 and illustrated with several applications to the design phase of experimental programs.

In the second part of the book, Chapters 5–7, we present additional considerations in the design of experiments and illustrate the use of uncertainty analysis in the debugging, execution, data analysis, and reporting phases of an experiment. The use of regression to obtain mathematical expressions that represent the data is discussed, along with a suggested technique for specifying the uncertainty that should be associated with the use of such curvefits.

The appendices provide statistical tables and a detailed development of the uncertainty propagation expression. Problems are provided at the end of each chapter to give practical exercises for the material presented.

This book was developed from notes used and revised by the authors over the past eight years in a junior/senior level mechanical engineering laboratory course, in a graduate course taken by master's and doctoral students in disciplines ranging from aerospace engineering to wood science, and in off-campus courses offered to practicing engineers and scientists. In the undergraduate laboratory course (which meets for three hours one afternoon each week for fourteen weeks), the material in Chapters 1–4 is introduced and selected ideas from Chapters 5, 6, and 7 are briefly discussed. This presentation is done in the context of planning, designing, and executing four experiments and analyzing and reporting the results. In a subsequent laboratory course, the students apply these techniques to additional experimental projects.

For the three credit hour graduate course, the material in the book (including Appendix B) is covered in depth and is supplemented with assigned readings from the literature and study of the ANSI/ASME Standard. Individual projects leading to a final report covering the planning and detailed analysis and design of some significant experimental program are a major focus of the course. This course has been taught quite successfully in a seminar format and in a traditional lecture-style format that was videotaped for use by off-campus students.

We believe this book will prove extremely useful to graduate students involved in research projects and to practicing engineers. The combination of narrative discussion, examples, and appendices should enable a graduate student or engineer to master the basic ideas of the application of uncertainty analysis to experimentation even without a formal course in the subject, and the references that are cited serve as a guide into the important literature on the subject.

Uncertainty analysis is an area that is still growing and being developed. Specific techniques in uncertainty analysis are still sometimes the subject of spirited debate. We have tried our best to distill what we have learned from coursework, the literature, and our own experience into a logical, consistent approach to experimentation using uncertainty analysis. In so doing, we have tried to give proper credit for ideas and techniques throughout the book— if we have failed to do this in any instance, it was truly unintentional. We invite the readers to communicate to us any comments or criticisms that would improve the explanation and application of the material presented. We also welcome suggestions for additional material that would enhance the ideas presented in the book.

HUGH W. COLEMAN
W. GLENN STEELE, JR.

Mississippi State University
August, 1989

EXPERIMENTATION AND UNCERTAINTY ANALYSIS FOR ENGINEERS

1

EXPERIMENTATION, ERRORS, AND UNCERTAINTY

When the word "experimentation" is encountered, most of us immediately envision someone in a laboratory "taking data." This idea has been fostered by the innumerable pictures in periodicals and on television of an engineer or scientist in a white lab coat writing on a clipboard while noting the readings on gauges or watching something happen in an impressive complexity of laboratory glassware. To some extent, the manner in which laboratory classes are typically implemented in university curricula also reinforces this idea. Students often encounter most instruction in experimentation as demonstration experiments that are already set up when the students walk into the laboratory. Data are often taken under the pressure of time, and much of the interpretation of the data and the reporting of results is spent on trying to rationalize what went wrong and what the results "would have shown if . . . ."

Experimentation Is Not Just Data-Taking. Any engineer or scientist who subscribes to the widely held but erroneous belief that experimentation is making measurements in the laboratory will be a failure as an experimentalist. The actual data-taking portion of a well-run experimental program generally constitutes a small percentage of the total time and effort expended.

In this book, we will examine and discuss the steps and techniques involved in a logical, thorough approach to the subject of experimentation.

1-1 EXPERIMENTATION

1-1.1 Why Is Experimentation Necessary?

Why are experiments necessary? Why do we need to study the subject of experimentation? The experiments run in science and engineering courses

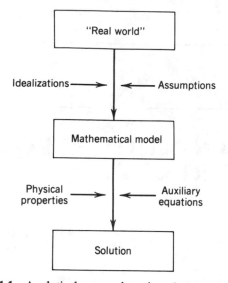

Figure 1.1   Analytical approach to the solution of a problem.

demonstrate physical principles and processes, but once these demonstrations are made and their lessons taken to heart, why bother with experiments? With the laws of physics we know, with the sophisticated analytical solution methods we study, with the increasing knowledge of numerical solution techniques, and with the awesome computing power available, is there any longer a need for experimentation in the "real world"?

These are fair questions to ask. To address them, it is instructive to consider the diagram in Figure 1.1, which illustrates a typical *analytical* approach to finding a solution to a physical problem. Experimental information is almost always required at one or more stages of the solution process, even when an analytical approach is used. Sometimes experimental results are necessary before realistic assumptions and idealizations can be made so that a mathematical model of the "real world" process can be formulated. In addition, experimentally determined information is generally present in the form of physical property values and the auxiliary equations (equations of state, etc.) necessary for obtaining a solution. So we see that even in situations in which the solution approach is analytical (or numerical), information from experiments is included in the solution process.

From a more general perspective, experimentation lies at the very foundations of science and engineering. *Webster's* [1] defines science as "systematized knowledge derived from observation, study, and experimentation carried on in order to determine the nature or principles of what is being studied." In discussing the scientific method, Shortley and Williams [2] state

The scientific method is the systematic attempt to construct theories that correlate wide groups of observed facts and are capable of predicting the results of future observations. Such theories are tested by controlled experimentation and are accepted only so long as they are consistent with all observed facts.

In many systems and processes of scientific and engineering interest, the geometry, boundary conditions, and physical phenomena are so complex that it is beyond our present technical capability to formulate satisfactory analytical or numerical models and approaches. In these cases, experimentation is necessary to define the behavior of the systems and/or processes (that is, to find a solution to the problem).

1-1.2 "Degree of Goodness" and Uncertainty Analysis

If we are using property data or other experimentally determined information in an analytical solution, we should certainly consider how "good" the experimental information is. Likewise, anyone comparing results of a mathematical model with experimental data (and perhaps also with the results of other mathematical models) should certainly consider the "degree of goodness" of the data when drawing conclusions based on the comparisons. This situation is illustrated in Figure 1.2. In Figure 1.2a the results of two different mathematical models are compared with each other and with a set of experimental data. The authors of the two models might have a fine time arguing over which model compares best with the data. In Figure 1.2b the same information is presented, but a range representing the likely amount of error in the experimental data has been plotted for each data point.

It should be immediately clear that once the "degree of goodness" of the data is taken into consideration, it is fruitless to argue for the validity of one model over another based only on how well the results match the data. From this example, one might conclude that even those with no ambitions of becoming experimentalists need an appreciation of the experimental process and the factors that influence the "degree of goodness" of experimental data.

Whenever the experimental approach is to be used to answer a question or to find the solution to a problem, the question of "how good" the results will be should be considered long before an experimental apparatus is constructed and data are taken. If the answer or solution must be known within 5%, say, for it to be useful to us, it would make no sense to spend the time and money to perform the experiment only to find that the likely amount of error in the results was considerably more than 5%.

From the preceding discussion, it is evident that the "degree of goodness" of experimental results is an idea that has basic importance to the experimentalist and theoretician alike. In this book, we will use the concept of uncertainty to describe the "degree of goodness" of a measurement or experimental result. Schenck [3] quotes S. J. Kline as defining uncertainty as "what we think the error would be if we could and did measure it by calibration." Uncertainty is thus an estimate of experimental error.

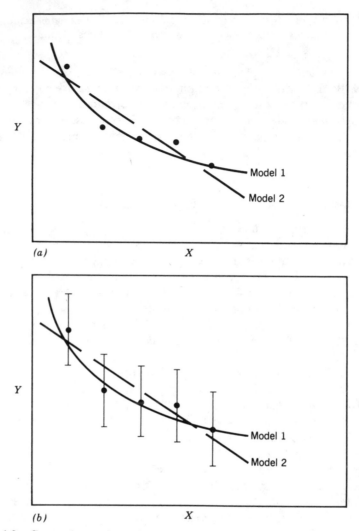

Figure 1.2 Comparison of mathematical model results with experimental data: (a) without error bars; (b) with error bars.

Uncertainty analysis (the analysis of the uncertainties in experimental measurements and results) is a powerful tool, particularly when it is used in the planning and design of experiments. As we will see in Chapter 3, there are realistic, practical cases in which all the measurements in an experiment can be made with 1% uncertainty, yet the uncertainty in the final experimental result will be greater than 50%. Uncertainty analysis, when used in the initial planning phase of an experiment, can identify such situations and save the experimentalist much time, money, and embarrassment.

1-2 THE EXPERIMENTAL APPROACH

1-2.1 Questions to Be Considered

When an experimental approach is to be used to find a solution to a problem, there are many questions that must be considered. Among these are

1. What question are we trying to answer? (What is the problem?)
2. How accurately do we need to know the answer? (How is the answer to be used?)
3. What physical principles are involved? (What physical laws govern the situation?)
4. What experiment or set of experiments might provide the answer?
5. What variables must be controlled? How well?
6. What quantities must be measured? How accurately?
7. What instrumentation is to be used?
8. How are the data to be acquired, conditioned, and stored?
9. How many data points must be taken? In what order?
10. Can the requirements be satisfied within the budget and time constraints?
11. What techniques of data analysis should be used?
12. What is the most effective and revealing way to present the data?
13. What unanticipated questions are raised by the data?
14. In what manner should the data and results be reported?

This is by no means an all-inclusive list, but it does indicate the range of factors that must be considered by the experimentalist. This might seem to be a discouraging and somewhat overwhelming list, but it need not be. With the aid of uncertainty analysis and a logical, thorough approach in each phase of an experimental program, the apparent complexities often can be reduced and the chances of achieving a successful conclusion enhanced.

A key point is to avoid becoming so immersed in the many details that must be considered that the overall objective of the experiment is forgotten. This may sound trite, but it is true nonetheless. We perform an experiment to find the answer to a question. We need to know the answer within some uncertainty, the magnitude of which is usually determined by the intended use of the answer. Uncertainty analysis is a tool that we use in making decisions in each phase of the experiment, always keeping in mind the desired result and uncertainty. Properly applied, this approach will guide us past the pitfalls that are usually not at all obvious and will enable us to obtain an answer with an acceptable uncertainty.

1-2.2 Phases of an Experimental Program

There are numerous ways that a general experimental program can be divided into different components or phases. For our discussions in this book, we will consider the experimental phases as planning, design, construction, debugging, execution, data analysis, and reporting of results. There are not sharp divisions between these phases—in fact, there is generally overlap and sometimes several phases will be ongoing simultaneously (as when something discovered during debugging leads to a design change and additional construction on the apparatus).

In the planning phase, we consider and evaluate the various approaches that might be used to find an answer to the question being addressed. This is sometimes referred to as the preliminary design phase.

In the design phase, we use the information found in the planning phase to specify the instrumentation needed and the details of the configuration of the experimental apparatus. The test plan is identified and decisions made on the ranges of conditions to be run, the data to be taken, the order in which the runs will be made, etc.

During the construction phase, the individual components are assembled into the overall experimental apparatus and necessary instrument calibrations performed.

In the debugging phase, the initial runs using the apparatus are made and the unanticipated problems (which must always be expected!) are addressed. Often, results obtained in the debugging phase will lead to some redesign and changes in the construction and/or operation of the experimental apparatus. At the completion of the debugging phase, the experimentalist should be confident that the operation of the apparatus and the factors influencing the uncertainty in the results are well understood.

During the execution phase, the experimental runs are made and the data acquired, recorded, and stored. Often the operation of the apparatus is monitored using checks that were designed into the system to guard against unnoticed and unwanted changes in the apparatus or operating conditions.

During the data analysis phase, the data are analyzed to determine the answer to the original question or the solution to the problem being investigated.

In the reporting phase, the data and conclusions should be presented in a form that will maximize the usefulness of the experimental results.

In the chapters that follow, we discuss a logical approach for each of these phases. We will find that the use of uncertainty analysis and related techniques (balance checks, for example) will help to ensure a maximum return for the time, effort, and financial resources invested.

1-3 BASIC CONCEPTS AND DEFINITIONS

There is no such thing as a perfect measurement. All measurements of a variable contain inaccuracies. Because it is important to have an understanding of these inaccuracies if we are to perform experiments (use the experimental approach to answer a question) or if we are simply to use values that have been experimentally determined, we must carefully define the concepts involved.

Unfortunately, there has been no consistent usage of nomenclature or definition of concepts concerning measurement uncertainty and uncertainty analysis. This has been partly the result of disagreements in the technical community over the concepts themselves and over the method of application of uncertainty analysis. Accuracy, fixed error, bias, precision, random error, uncertainty, confidence level, and related words have had somewhat different meanings in different presentations on the subject. This has been the case in widely used texts (Schenck [3], Taylor [4], and Holman [5], for example) as well as in the technical literature [Refs. 6–13, for example].

In our discussions, we will attempt to remain consistent with the basic nomenclature and definitions as set forth in the ANSI/ASME Standard on *Measurement Uncertainty* [14] issued in 1986. The authors feel that adherence by the technical community (insofar as practical) to the definitions and nomenclature in the Standard will remove some of the ambiguity from discussions of uncertainty analysis and thus have a positive long-term effect.

1-3.1 Fixed and Random Errors; Uncertainty

We will take the word *accuracy* to refer to the closeness of agreement between a measured value and the true value. The degree of inaccuracy or the total measurement error (δ) is the difference between the measured value and the true value. As shown in Figure 1.3a, the *total error* is the sum of the *bias error* and the *precision error*. The bias error (β) is the fixed, systematic, or constant component of the total error and is sometimes referred to simply as the bias. The precision error (ϵ) is the random component of the total error and is sometimes called the repeatability or repeatability error.

Suppose we are making a number of measurements, one after the other, of the value of a variable X that is absolutely steady. The k and $k + 1$ measurements are shown in Figure 1.3b. Since the bias is a fixed error, it is the same for each measurement. However, the precision error is a random error and will have a different value for each measurement. It then follows that the total error in each measurement will be different, since

$$\delta_i = \beta + \epsilon_i \qquad (1.1)$$

If we continued to take measurements as previously described until we had a sample of N readings, more than likely as N approached infinity the data

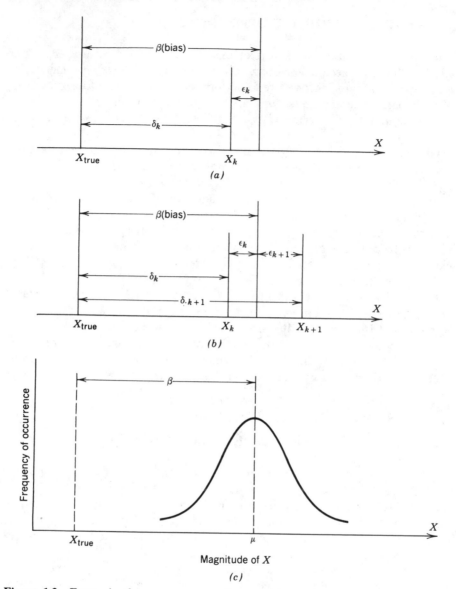

Figure 1.3 Errors in the measurement of a variable X: (a) single reading; (b) two readings; (c) infinite number of readings.

would behave as shown in Figure 1.3c. The bias would be given by the difference between the mean (average) value of the N readings, μ, and the true value of X, whereas the precision errors would cause the frequency of occurrence of the readings to be distributed about the mean value.

An example of this behavior is shown in Figure 1.4. A thermometer immersed in an insulated beaker of fluid was read independently by 24

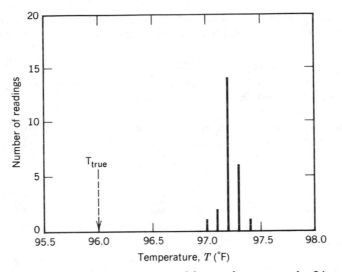

Figure 1.4 Histogram of temperatures read from a thermometer by 24 students.

students to the nearest one-tenth of a degree F. Unknown to the students, the thermometer was biased by a little over 1°F, and the true temperature of the fluid was 96.0°F. The temperatures read by the students are distributed around an average value of about 97.2°F and are offset (biased) from the true value of 96.0°F.

We will discuss this situation from a statistical viewpoint in the next chapter. For the moment, consider the broad implications of this type of behavior on our ability to specify what the error is in a measurement of X. Unfortunately, we do not know the true value of X. It is therefore not possible to specify what the exact bias and precision errors are in a given measurement of X.

If we want to make a statement about the value of X based on our measurements, about the best we are able to do is to say that we are $C\%$ confident that the true value of X lies within the interval

$$X_{best} \pm U_X$$

X_{best} is usually assumed to be the mean value of the N readings we have taken (or *the* reading if $N = 1$), and U_X is the uncertainty in X that corresponds to our estimate (with $C\%$ confidence) of the combination of bias and precision error.

The confidence specification is necessary because we have made an estimate. We can always be 100% confident that the true value of some quantity will lie between plus and minus infinity, but specifying U_X as infinite provides no useful information to anyone. It is not necessary to perform an experiment to find that result!

The idea of degree of confidence in an uncertainty specification is vividly and humorously illustrated in an anecdote reported by Abernethy et al. [9]:

> In the 1930's, P. H. Myers at NBS and his colleagues were studying the specific heat of ammonia. After several years of hard work, they finally arrived at a value and reported the result in a paper. Toward the end of the paper, Myers declared: "We think our reported value is good to one part in 10,000; we are willing to bet our own money at even odds that it is correct to two parts in 10,000; furthermore, if by any chance our value is shown to be in error by more than one part in 1000, we are prepared to eat our apparatus and drink the ammonia!"

Sometimes the uncertainty specification must correspond to more than an estimate of the "goodness" with which we can measure something. This is true for cases in which the quantity of interest has a variability unrelated to the bias and precision errors inherent in the measurement system. When discussing this, it is helpful to consider two broad categories—timewise experiments and sample-to-sample experiments.

In a typical timewise experiment we might want to measure a fluid flow rate in a system operating at a steady-state condition. Although the system might be in "steady" operation, there inevitably will be some time variations of flow rate that will appear as precision errors in a series of flow rate measurements taken over a period of time. In addition, inability to reset the system at exactly the same operating condition from trial to trial will cause additional data scatter.

In sample-to-sample experiments, measurements are made on multiple samples so that in a sense, sample identity corresponds to the passage of time in timewise experiments. In sample-to-sample experiments, the variability inherent in the samples themselves causes variations in measured values in addition to the precision errors in the measurement system.

As a practical example of such a case, suppose we have been retained to determine the heating value of lignite in a particular geographic region. Lignite is a soft, brown coal with relatively low heating value and high ash and moisture content. However, a utility company is interested in the possibility of locating a new lignite-fueled power plant in the midst of a region (shown in Figure 1.5) that contains several large deposits of lignite. The savings in coal transportation costs might offset the negative effects of the low heating value of lignite and make such a power plant economically feasible. A decision on the economic feasibility depends critically, of course, on the lignite heating value used in the calculations.

The heating value of a 1-g sample of lignite can be determined with an uncertainty of less than 2% in the laboratory using a bomb calorimeter. However, because of large variations in the composition of the lignite within a single deposit and from deposit to deposit, the heating values determined for samples from the northern and southern portions of Deposit C, for instance, might differ by 20% or more. Even larger differences might very well be

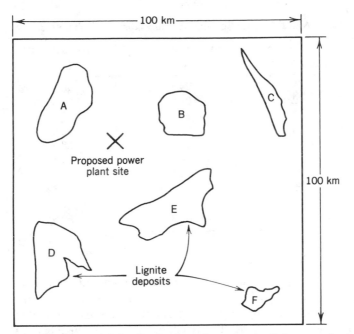

Figure 1.5 Region containing lignite deposits and potential power plant site.

observed when results are compared for samples from entirely different deposits.

In this situation, the variation within a set of individual measurements of lignite heating value is not primarily due to errors in the measurements. It is due to the natural variation of the physical variable that is being measured. The amount of the variation within a set of measurements will differ depending on whether the set is taken from within a small area of a single deposit, from the entire area of a single deposit, or from different deposits. We might conclude from this example that the answer to the question "What is the uncertainty in the lignite heating value?" depends on what is meant when one says "the lignite." The answer will be one thing for a single 1-g sample and quite another thing for Deposits A, B, and E considered together.

1-3.2 Bias Errors and Calibration

To illustrate what can be achieved toward the reduction of bias errors by calibration, consider that we have two voltmeters, A and B, that we wish to calibrate at 100 V dc. To provide the reference 100 V, a standard voltage source is available that has an extremely stable output (negligible precision error) and an "absolute accuracy" guaranteed by the manufacturer of 0.5 V.

Multiple readings of the 100.0-V output of the standard voltage source are taken with both meters. The first 10 readings for each meter are shown in

TABLE 1.1   Calibration Readings for Two Voltmeters

Reading Number	Voltmeter A (V)	Voltmeter B (V)
1	104.5	90.0
2	101.5	91.5
3	96.0	89.5
4	105.5	90.5
5	97.0	88.5
6	100.0	89.5
7	95.5	90.5
8	103.5	89.5
9	101.0	91.0
10	101.5	89.5

Table 1.1 and are plotted in Figure 1.6. After many additional readings, it is found that the average of the readings is 100.6 V for meter A and 90.0 V for meter B, and these averages remain essentially the same no matter how many additional readings are made. The fixed, constant difference between the input and the average of the readings is a bias error. This is +0.6 V for meter A and −10.0 V for meter B. Thus, we can correct for these systematic errors by subtracting 0.6 V from any readings taken with A or adding 10.0 V to any readings taken with B. These values, of course, are valid only for an input value close to 100.0 V. The bias error for an input of 150 V, for instance, could very well be different and would have to be determined by calibration at 150 V.

It is tempting, after considering examples such as this, to conclude that we can correct for or eliminate bias errors by calibration. In fact, bias errors are often summarily dismissed in books and articles on error analysis or uncertainty analysis by a simple assumption that "all bias errors have been eliminated by calibration." If we consider the voltmeter example more closely, however, we realize that the "absolute accuracy" of 0.5 V that the manufacturer specified for the standard source means that the minimum bias that we can assume remains after calibration of the voltmeters is ±0.5 V. From this example, we conclude that some bias errors can be eliminated by calibration, but only to the limit of the bias error associated with the standard used in the calibration procedure.

There are other forms of bias errors caused by things such as transducer installation effects and environmental effects that will be discussed in some detail in later chapters. An example is the case of a thermocouple used to measure the temperature of a hot gas in an exhaust pipe whose wall is at a lower temperature than the gas. The loss of energy by thermal radiation from the thermocouple to the pipe walls causes the thermocouple to be at a temperature lower than the gas temperature. This bias is always in the same

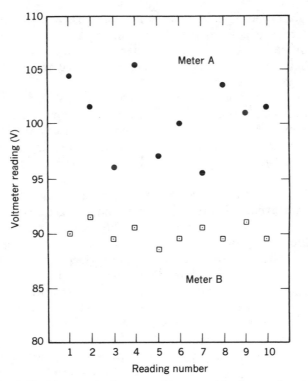

Figure 1.6   Calibration results for two voltmeters with a known input of 100 V.

direction, regardless of which particular thermocouple is used. In such cases, corrections are usually made during data reduction to try to minimize the effect of the bias.

1-3.3   Repetition and Replication

In discussing biases, precision errors, and uncertainties in measurements, it is useful to draw a careful distinction between the meanings of the words "repetition" and "replication." We will use repetition in its common sense, that is, to mean that something is repeated. When we use the word replication, we will be specifying that the repetition is carried out in a very specific manner. The reason for doing this is that different factors influence the errors in a series of measurements depending on how the repetition is done. The idea of considering uncertainties at different orders of replication level was suggested by Moffat [7, 11, 15] for the timewise category of experiments, and we will find it useful also for sample-to-sample experiments. As did Moffat, we will find it convenient to define three different levels of replication—zeroth order, first order, and Nth order.

At the zeroth-order replication level in a timewise experiment, the process being measured is hypothesized to be absolutely steady. This allows only the variations inherent in the measuring system itself to contribute to precision errors. In a sample-to-sample type of experiment, this corresponds to consideration of a single fixed sample.

If we make a zeroth-order estimate of the bias and precision errors associated with an instrument, this is the "best" we would ever be able to achieve if we used that instrument. If this zeroth order estimate shows that the bias and/or precision errors are larger than those that are allowable in our experiment, then it is obvious we should not perform the experiment using that instrument.

In considering uncertainties at the first-order replication level in a timewise experiment, we hypothesize that time runs but all instrument identities are fixed. At this level of replication, the variability of the experimental measurements is influenced by all of the factors that contribute to unsteadiness during repeated trials with the experimental apparatus. Depending on the length of time covered by a replication at the first-order level, different factors such as variations of humidity, barometric pressure, ambient temperature, and the like can influence the random portion of the experimental uncertainty. In addition, our inability to exactly reproduce a set point with the experimental apparatus influences the variations in experimental results.

A first-order estimate of the precision errors shows the "scatter" that we would expect to observe during the course of a timewise experiment. If on repeated trials over the course of time we observe a scatter significantly larger than that given by the first-order estimate, it is an indication that there is some effect that has not been taken into account and that we should probably investigate further. This use of first-order precision error estimates in debugging timewise experiments is discussed in detail in Chapter 6.

In considering uncertainties at the first-order replication level in a sample-to-sample experiment, we imagine all instruments remain the same as sample after sample is tested. Additional variations observed above the level of precision errors at the zeroth order are indicative of the variations in the samples themselves.

When considering what uncertainties should be specified when we speak of where the true value lies relative to our measurements, we make estimates at the Nth-order replication level. Such Nth-order estimates of uncertainty include the first-order replication level estimates of precision errors together with all of the bias errors which influence our measurements.

Using Moffat's concept for timewise experiments, at the Nth-order replication level both time and instrument identities are considered to vary. At this level, for each reading each instrument is considered to have been replaced by another of the same type. What this essentially means is that the bias error associated with a particular instrument becomes a random variable (similar to precision error) when the instrument identity is allowed to vary. To see this, consider that a particular model of pressure gauge might be specified by the manufacturer as "accurate within plus or minus one percent of full scale." The particular gauge we use might actually read high by 0.75% of full scale (a bias error). If we replace that gauge by another "identical" gauge, it might very well read low by 0.37% of full scale. Thus, when we consider replacing each instrument with an "identical" instrument for each reading, the bias error associated with the instrument becomes a random error at the Nth-order level of replication.

From the preceding discussion, it should be evident that the reason for considering different levels of replication is that the factors that are considered to influence the uncertainty in the experiment differ with the level of replication. We shall make use of this point as we consider the analysis and design of experiments in later chapters.

REFERENCES

1. *Webster's New Twentieth Century Dictionary*, 2nd ed., Simon & Schuster, 1979.

2. Shortley, G., and Williams, D., *Elements of Physics*, 4th ed., Prentice-Hall, Englewood Cliffs, NJ, 1965.

3. Schenck, H., *Theories of Engineering Experimentation*, 3rd ed., McGraw-Hill, New York, 1979.

4. Taylor, J. R., *An Introduction to Error Analysis: The Study of Uncertainties in Physical Measurements*, University Science Books, Mill Valley, CA, 1982.

5. Holman, J. P., *Experimental Methods for Engineers*, 4th ed., McGraw-Hill, New York, 1984.

6. Kline, S. J., and McClintock, F. A., "Describing Uncertainties in Single-Sample Experiments," *Mechanical Engineering*, Vol 75, Jan. 1953, pp 3–8.

7. Moffat, R. J., "Contributions to the Theory of Single-Sample Uncertainty Analysis," *J. Fluids Engineering*, Vol 104, June 1982, pp 250–260.

8. Kline, S. J., "The Purposes of Uncertainty Analysis," *J. Fluids Engineering*, Vol 107, June 1985, pp 153–160.

9. Abernethy, R. B., Benedict, R. P., and Dowdell, R. B., "ASME Measurement Uncertainty," *J. Fluids Engineering*, Vol 107, June 1985, pp 161–164.

10. Smith, R. E., Jr., and Wehofer, S., "From Measurement Uncertainty to Measurement Communications, Credibility, and Cost Control in Propulsion Ground Test Facilities," *J. Fluids Engineering*, Vol 107, June 1985, pp 165–172.

11. Moffat, R. J., "Using Uncertainty Analysis in the Planning of an Experiment," *J. Fluids Engineering*, Vol 107, June 1985, pp 173–178.

12. Lassahn, G. D., "Uncertainty Definition," *J. Fluids Engineering*, Vol 107, June 1985, p 179.

13. Kline, S. J., "1983 Symposium on Uncertainty Analysis Closure," *J. Fluids Engineering*, Vol 107, June 1985, pp 181–182.

14. *Measurement Uncertainty*, ANSI/ASME PTC 19.1-1985 Part 1, 1986. (Available from ASME Order Dept., 22 Law Drive, Box 2300, Fairfield, New Jersey 07007-2300.)

15. Moffat, R. J., "Describing the Uncertainties in Experimental Results," *Experimental Thermal and Fluid Science*, Vol 1, Jan. 1988, pp 3–17.

CHAPTER 1

Problems

1.1 Consider the lignite heating value example discussed in Section 1-3.1. A salesman drops by your office and says he can supply a calorimetry system that can determine the heating value of a 1-g sample of lignite with an uncertainty of about 1% as opposed to the 2% with the present system. Is this idea worth considering if you need to find the heating value of the lignite in Deposit E?

1.2 Consider the two voltmeters for which calibration results are presented in Table 1.1 and Figure 1.6. If you needed to measure a voltage that was about 100 V, which of the calibrated meters would you choose—Meter A or Meter B? Why?

1.3 A Pitot-static probe is mounted to monitor the exit velocity of air exhausting from a duct in a process unit. The differential pressure output from the probe is applied to an analog pressure gauge, and a technician reads the gauge and notes the reading every hour. The operating condition of the process unit is held as steady as possible by its control system. If the gauge is located in a room with a conditioned environment, list the possible factors involved in causing "scatter" in the readings. What additional factors are involved if the gauge is located in a room where the temperature and humidity are uncontrolled? If the process unit is shut down and then restarted and reset to the same operating condition? If the gauge is replaced by another of the same model number from the same manufacturer?

2

STATISTICAL CONSIDERATIONS IN MEASUREMENT UNCERTAINTIES

In the previous chapter, the concepts of bias and precision errors were discussed and the idea of the degree of confidence in an error estimate was introduced. Since precision errors are random errors, it is necessary to consider how we can determine some quantitative value of the "scatter" in the measurements of a variable and the degree of confidence to associate with such a value. To develop such a quantitative measure, we must consider the statistics of random measurement errors. Statistics is the study of mathematical properties of distributions of numbers (such as the multiple readings of voltmeter output for a constant voltage input discussed in Chapter 1).

2-1 STATISTICAL DISTRIBUTIONS

Suppose that the output (an analog voltage) of a pressure transducer is monitored over a period of time while the transducer is supposedly measuring a "constant" input pressure. The readings that are taken might appear as shown in the histogram plotted in Figure 2.1. The readings scatter about a central value of about 86 mV, with some readings being higher and some lower. If additional readings were taken so that the total number of readings approached infinity, the resulting histogram curve would become smoother and more "bell-shaped" as shown in Figure 2.2.

The distribution of readings defined if an infinite number of samples could be taken is called the *parent distribution* or *parent population*. In reality, of course, we never have the time or resources to obtain an infinite number of readings, so we must work with a *sample distribution* composed of a finite

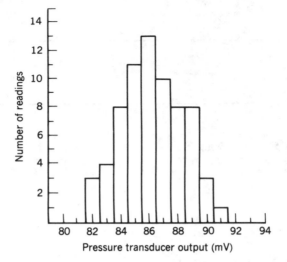

Figure 2.1   Histogram of readings of output of a pressure transducer.

number of readings taken from the parent population. In the following sections, we will first discuss the statistical characteristics of Gaussian parent distributions and then of sample distributions that may contain relatively large or relatively small numbers of readings. In addition, we will consider how the concepts developed for precision errors can be extended to estimates of bias errors. Finally, in Section 2-5 the combination of estimates of precision and bias errors into an overall uncertainty in a measurement is considered.

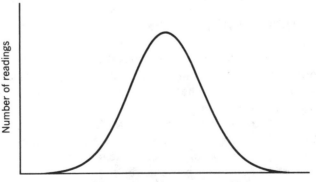

Figure 2.2   Distribution of readings of output of a pressure transducer as the number of readings approaches infinity.

2-2   THE GAUSSIAN DISTRIBUTION

For cases in which the variation in the readings results from a combination of many small errors of equal magnitude with each of the errors being equally likely to be positive as negative, the smooth distribution of an infinite number of readings coincides with the Gaussian, or normal, distribution. The Gaussian distribution has been found to describe more real cases of experimental and instrument variability than any other distribution and is the one assumed in the approach recommended in the Standard [1]. We will assume that the Gaussian distribution applies in the cases that we will be considering in this book. The effect of this assumption when error distributions are in fact non-Gaussian will be discussed in Section 2-3.4.

2-2.1   Mathematical Description

The equation for the Gaussian distribution is

$$f(X) = \frac{1}{\sigma\sqrt{2\pi}} e^{-(X-\mu)^2/2\sigma^2} \tag{2.1}$$

where $f(X)\,dX$ is the probability that a single measurement of X will lie between X and $X + dX$, μ is the distribution mean defined by

$$\mu = \lim_{N\to\infty} \frac{1}{N} \sum_{i=1}^{N} X_i \tag{2.2}$$

and σ is the distribution standard deviation defined by

$$\sigma = \lim_{N\to\infty} \left[\frac{1}{N} \sum_{i=1}^{N} (X_i - \mu)^2 \right]^{1/2} \tag{2.3}$$

The square of the standard deviation is known as the variance of the distribution.

A plot of Eq. (2.1) is shown in Figure 2.3 for two cases in which the mean μ is equal to 5.0 and the standard deviation is equal to 0.5 and 1.0, respectively. As the value of σ increases, the range of values of X expected also increases. Larger values of σ, therefore, correspond to cases in which the "scatter" in the X readings is large and thus the precision error is large.

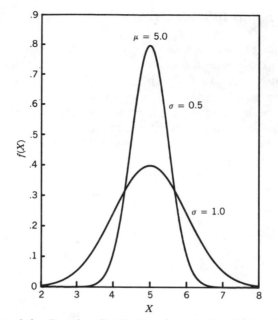

Figure 2.3   Plot of the Gaussian distribution showing the effect of different values of standard deviation.

Equation (2.1) is in a normalized form such that

$$\int_{-\infty}^{\infty} f(X) \, dX = 1.0 \tag{2.4}$$

This makes intuitive sense, since we would expect that the probability of a reading being between plus and minus infinity must equal one.

Now suppose that we wish to determine the probability that a reading from a Gaussian parent population will fall within a specified range, say plus or minus ΔX, about the mean value μ. This can be expressed as

$$\text{Prob}(\Delta X) = \int_{\mu - \Delta X}^{\mu + \Delta X} \frac{1}{\sigma \sqrt{2\pi}} e^{-(X-\mu)^2 / 2\sigma^2} \, dX \tag{2.5}$$

This integral cannot be evaluated in a closed form, and if its value was tabulated for a range of ΔX values, there would have to be a table for every pair of (μ, σ) values. Rather than attempt to generate an infinite number of tables, it is more logical to normalize the integral so that only a single table is required.

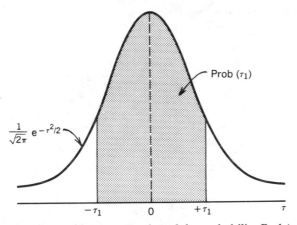

Figure 2.4   Graphic representation of the probability Prob(τ_1).

If a normalized deviation from the mean value μ is defined as

$$\tau = \frac{X - \mu}{\sigma} \tag{2.6}$$

then Eq. (2.5) can be rewritten as

$$\text{Prob}(\tau_1) = \frac{1}{\sqrt{2\pi}} \int_{-\tau_1}^{\tau_1} e^{-\tau^2/2} \, d\tau \tag{2.7}$$

where $\tau_1 = \Delta X/\sigma$.

The value of Prob(τ_1) corresponds to the area under the Gaussian curve between $-\tau_1$ and $+\tau_1$ as shown in Figure 2.4. Prob(τ) is called the "two-tailed" probability since both the negative and positive tails of the distribution are included in the integration. Since the Gaussian distribution is symmetric, the probability of a reading with a nondimensional deviation between 0 and τ or between $-\tau$ and 0 is (1/2)Prob(τ) and is called the "single-tailed" probability. Values of Prob(τ) for τ varying from 0.0 to 5.0 are presented in Table A.1 of Appendix A.

Example 2.1

A Gaussian distribution has a mean μ of 5.00 and a standard deviation σ of 1.00. Find the probability of a single reading from this distribution being

a. Between 4.50 and 5.50
b. Between 4.50 and 5.75

c. Equal to or less than 6.50
d. Between 6.00 and 7.00.

a. For $\mu = 5.0$ and $\sigma = 1.0$, want Prob$(4.5 < X_i < 5.5)$; let $X_1 = 4.50$ and $X_2 = 5.50$, then

$$\tau_1 = \frac{X_1 - \mu}{\sigma} = \frac{4.50 - 5.00}{1.00} = -0.50$$

$$\tau_2 = \frac{X_2 - \mu}{\sigma} = \frac{5.50 - 5.00}{1.00} = +0.50$$

Since $\tau_1 = \tau_2$, this is exactly what is tabulated in the two-tailed probability tables. Therefore Prob = Prob(0.5) (see Figure 2.5). From Table A.1,

$$\text{Prob}(\tau) = \text{Prob}(0.5) = 0.3829 \approx 38.3\%$$

b. Want Prob$(4.50 < X_i < 5.75)$; let $X_1 = 4.50$ and $X_2 = 5.75$; from (a) $\tau_1 = -0.50$

$$\tau_2 = \frac{X_2 - \mu}{\sigma} = \frac{5.75 - 5.00}{1.00} = +0.75 \qquad \text{(see Figure 2.6)}$$

Prob $= \frac{1}{2}$ Prob$(0.5) + \frac{1}{2}$ Prob(0.75)

Prob $= \frac{1}{2}(0.3829) + \frac{1}{2}(0.5467) = 0.4648 \approx 46.5\%$

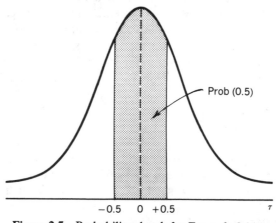

Figure 2.5 Probability sketch for Example 2.1(a).

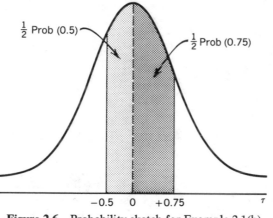

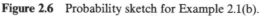

$\frac{1}{2}$ Prob (0.5)

$\frac{1}{2}$ Prob (0.75)

−0.5   0   +0.75 τ

Figure 2.6   Probability sketch for Example 2.1(b).

c. Want Prob($X_i \leq 6.50$); let

$$\tau_1 = \frac{X_i - \mu}{\sigma} = \frac{6.5 - 5.0}{1.0} = 1.50 \qquad \text{(see Figure 2.7)}$$

Prob $= 0.5 + \frac{1}{2}$ Prob$(1.50) = 0.5 + 0.8664/2$

Prob $= 0.9332 \approx 93.3\%$

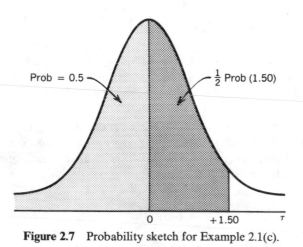

Prob $= 0.5$

$\frac{1}{2}$ Prob (1.50)

0   +1.50 τ

Figure 2.7   Probability sketch for Example 2.1(c).

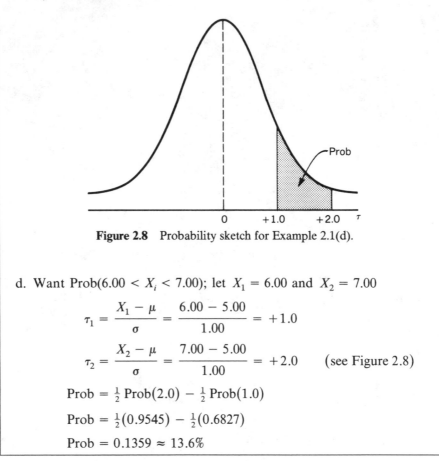

Figure 2.8   Probability sketch for Example 2.1(d).

d. Want Prob(6.00 < X_i < 7.00); let X_1 = 6.00 and X_2 = 7.00

$$\tau_1 = \frac{X_1 - \mu}{\sigma} = \frac{6.00 - 5.00}{1.00} = +1.0$$

$$\tau_2 = \frac{X_2 - \mu}{\sigma} = \frac{7.00 - 5.00}{1.00} = +2.0 \qquad \text{(see Figure 2.8)}$$

$$\text{Prob} = \tfrac{1}{2} \text{Prob}(2.0) - \tfrac{1}{2} \text{Prob}(1.0)$$

$$\text{Prob} = \tfrac{1}{2}(0.9545) - \tfrac{1}{2}(0.6827)$$

$$\text{Prob} = 0.1359 \approx 13.6\%$$

2-2.2   Confidence Intervals in a Gaussian Distribution

Based on what we have seen in the previous example and on the probabilities tabulated in Table A.1 of Appendix A, we can observe that 50% of the readings from a Gaussian parent population are within $\pm 0.675\sigma$ of the mean, 68.3% are within $\pm 1.0\sigma$ of the mean, 95% are within $\pm 1.96\sigma$ of the mean, 99.7% are within $\pm 3.0\sigma$ of the mean, and 99.99% are within $\pm 4.0\sigma$ of the mean.

Suppose that we had a Gaussian parent population of readings X with a mean value μ and a standard deviation σ. If we were to take one more reading X_i, within what interval could we be 95% confident the reading would fall?

Using Eq. (2.6), we can define the normalized deviation of X_i from μ as

$$\tau = \frac{X_i - \mu}{\sigma}$$

and we can obtain from Table A.1 that for Prob(τ) = 0.95, τ = 1.96. This probability expression can be written as

$$\text{Prob}\left(-1.96 \leq \frac{X_i - \mu}{\sigma} \leq 1.96\right) = 0.95 \qquad (2.8)$$

or, after rearranging to isolate X_i,

$$\text{Prob}(\mu - 1.96\sigma \leq X_i \leq \mu + 1.96\sigma) = 0.95 \qquad (2.9)$$

Thus, knowing that 95% of the population lies within $\pm 1.96\sigma$ of the mean μ, we can be 95% confident that a single reading will fall within this $\pm 1.96\sigma$ interval about the mean. Stated another way, $+1.96\sigma$ and -1.96σ are the upper and lower limits on the 95% confidence interval for a single reading of X. We can also speak of $\pm 1.96\sigma$ as the 95% confidence limits.

This concept of a confidence interval is fundamental to uncertainty analysis in experimentation. For instance, from a test control point of view, we might want to know the range around a mean value in which acceptable readings could fall if they were from the same Gaussian parent population.

If we turned our point of view around 180°, we might ask within what interval about a single reading X_i would we expect the mean value of the distribution to lie at a confidence level of 95%? Upon rearranging Eq. (2.8) to isolate μ, we find

$$\text{Prob}(X_i - 1.96\sigma \leq \mu \leq X_i + 1.96\sigma) = 0.95 \qquad (2.10)$$

Thus, as shown in Figure 2.9, we can be 95% confident that the mean μ of the distribution will fall within $\pm 1.96\sigma$ of the single reading X_i. This conclusion

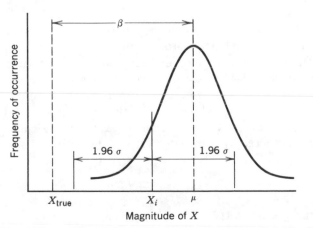

Figure 2.9 The 95% confidence interval about a single reading from a Gaussian distribution.

follows since 95% of the readings from the distribution will fall within $\pm 1.96\sigma$ of the mean μ. As we will see in the next section, the parent population mean value μ is usually not available. Therefore, this concept of a 95% confidence interval allows us to estimate the range that should contain μ.

2-3 SAMPLES FROM A GAUSSIAN PARENT POPULATION

The preceding discussion has assumed we were working with a Gaussian parent population, which would be well-described if we took an infinite number of samples. However, it is impractical to expect anything approaching an infinite number of samples in a realistic experimental situation. We must therefore consider the statistical properties of a sample population that consists of a finite number of samples drawn from a parent distribution.

2-3.1 Statistical Parameters of a Sample Population

The mean of the sample population is defined by

$$\overline{X} = \frac{1}{N} \sum_{i=1}^{N} X_i \tag{2.11}$$

where N is the number of individual readings X_i. The precision index of the sample population is defined by

$$S_X = \left[\frac{1}{N-1} \sum_{i=1}^{N} \left(X_i - \overline{X} \right)^2 \right]^{1/2} \tag{2.12}$$

where the $(N-1)$ occurs instead of N because the sample mean $\overline{X}$ is used instead of μ. This results in the loss of one degree of freedom since the same sample used to calculate S_X has already been used to calculate $\overline{X}$. S_X is sometimes called the sample standard deviation.

Another statistical parameter of interest is the precision index associated with the sample mean $\overline{X}$. Suppose that five sets of $N = 50$ readings are taken from a Gaussian parent population with mean μ and standard deviation σ and that a mean value is calculated for each of the five sets using Eq. (2.11). We would certainly not expect the five mean values to be the same. In fact, the sample means are normally distributed [2] with mean μ and standard deviation

$$\sigma_{\overline{X}} = \sigma / \sqrt{N} \tag{2.13}$$

The implications of this relationship are very important. One way to decrease the random component of the uncertainty in a measured value is to take many readings and average them. The inverse square root relationship of (2.13) indicates that this is a situation with rapidly diminishing returns—to reduce

$\sigma_{\bar{X}}$ by a factor of two, four times as many readings are required, whereas 100 times as many readings must be taken to reduce $\sigma_{\bar{X}}$ by a factor of ten.

Of course, the true standard deviation σ of the distribution is unknown, so in practice we must use the precision index of the mean, which is defined as

$$S_{\bar{X}} = S_X/\sqrt{N} \qquad (2.14)$$

where S_X is the precision index of the sample of N readings as given by Eq. (2.12). The precision index of the mean is sometimes called the sample standard deviation of the mean.

2-3.2 Confidence Intervals in Sample Populations

As we saw in Section 2-2.2, for a Gaussian distribution with mean μ and standard deviation σ

$$\text{Prob}(X_i - 1.96\sigma \leq \mu \leq X_i + 1.96\sigma) = 0.95 \qquad (2.10)$$

Thus we can say with 95% confidence that the mean μ of the parent distribution is within $\pm 1.96\sigma$ of a single reading from that distribution.

Recalling from the previous section that the mean $\bar{X}$ of a sample population of size N drawn from this same Gaussian distribution is itself normally distributed with standard deviation $\sigma/\sqrt{N}$, we can write

$$\text{Prob}\left(\bar{X} - 1.96 \frac{\sigma}{\sqrt{N}} \leq \mu \leq \bar{X} + 1.96 \frac{\sigma}{\sqrt{N}} \right) = 0.95 \qquad (2.15)$$

Thus, we can also say with 95% confidence that the mean μ of the parent distribution is within $\pm 1.96\sigma/\sqrt{N}$ of the sample mean $\bar{X}$ computed from N readings. The width of the 95% confidence interval in Eq. (2.15) is narrower than the one in Eq. (2.10) by a factor of $1/\sqrt{N}$.

Once again, the problem we face in actual experimental situations is that we do not know the value of σ—what we have is S_X, the precision index of a finite sample of N readings. S_X is only an estimate of the value of σ. Comparing Eqs. (2.12) and (2.3), we can see that the value of S_X approaches σ as the number of readings N in the sample approaches infinity.

If we are still interested in determining a 95% confidence interval, we are then forced to follow the same approach as shown in Eq. (2.8) and to seek the value of t that satisfies

$$\text{Prob}\left(-t \leq \frac{X - \mu}{S_X} \leq t \right) = 0.95 \qquad (2.16)$$

and

$$\text{Prob}\left(-t \leq \frac{\bar{X} - \mu}{S_X/\sqrt{N}} \leq t \right) = 0.95 \qquad (2.17)$$

where t is no longer equal to 1.96 because S_X is only an estimate of σ based on a finite number of readings N.

The variables

$$\frac{X - \mu}{S_X} \quad \text{and} \quad \frac{\overline{X} - \mu}{S_X/\sqrt{N}}$$

are not normally distributed—that is, their behavior is not Gaussian. Rather, they follow the t distribution with $N - 1$ degrees of freedom, ν, and the values of t that satisfy (2.16) and (2.17) are functions of the size N of the sample population. Table A.2 in Appendix A presents tabular values of t as a function of ν for several different confidence levels.

From Table A.2 we see that the 95% confidence interval is wider for smaller N: $t = 2.57$ for $N = 6$ as opposed to $t = 2.04$ for $N = 31$, for instance. The 95% confidence level value of t approaches the Gaussian value of 1.96 as N approaches infinity. In practice, for 95% confidence estimates the Standard [1] recommends the use of the value of t in the table for $N < 31$ and the value $t = 2.00$ for $N = 31$ or more. This in effect rounds off the 1.96 value of the Gaussian distribution to 2.00.

It is very convenient to define a quantity analogous to the 95% confidence limit that was discussed in relation to confidence intervals for a Gaussian distribution. We will define the 95% confidence limit for a sample of N measurements of X drawn from a Gaussian distribution as the precision limit, P, where

$$P_X = tS_X \tag{2.18}$$

and

$$P_{\overline{X}} = tS_{\overline{X}} = tS_X/\sqrt{N} \tag{2.19}$$

and where t is taken from Table A.2 if $N < 31$ and $t = 2.0$ for $N \geq 31$. When we consider sample populations drawn from a Gaussian parent population, the $\pm P$ interval will be the interval within which we expect μ to lie with 95% confidence. Upon rearranging Eqs. (2.16) and (2.17) to isolate μ, the confidence interval expressions become

$$\text{Prob}(X_i - P_X \leq \mu \leq X_i + P_X) = 0.95 \tag{2.20}$$

and

$$\text{Prob}(\overline{X} - P_{\overline{X}} \leq \mu \leq \overline{X} + P_{\overline{X}}) = 0.95 \tag{2.21}$$

Example 2.2

A thermometer immersed in an insulated beaker of fluid was read independently by 24 students to the nearest 0.1°F. The mean of the sample of 24 readings is $\overline{T} = 97.22°F$ and the precision index of the sample is $S_T =$

0.082°F as calculated using Eq. (2.12). Find the precision limits for this case and discuss their meaning.

The precision limits for this case are

$$P_T = tS_T$$

and

$$P_{\bar{T}} = tS_{\bar{T}} = tS_T/\sqrt{N}$$

For $N = 24$, the number of degrees of freedom ν is 23 and we find

$$t = 2.069$$

from Table A.2 in Appendix A. Using this value

$$P_T = (2.069)(0.082°F) = 0.170°F$$

$$P_{\bar{T}} = (2.069)(0.082°F)/\sqrt{24} = 0.0346°F$$

The interval defined by

$$\bar{T} \pm P_T = 97.2°F \pm 0.2°F$$

gives the range within which we expect, with 95% confidence, the next reading to lie if another one is taken. (This range also will contain μ_T 95 times out of 100.) Note that we have used a common sense approach in rounding off the final numbers to the nearest tenth of a degree.

The interval defined by

$$\bar{T} \pm P_{\bar{T}} = 97.22°F \pm 0.03°F$$

gives the range within which we expect (with 95% confidence) the mean value μ_T of the parent population to fall. Referring back to Figure 2.9, we note that μ_T is the *biased* mean value and would only correspond to the true temperature T_{true} if the bias in the temperature readings was zero. Note again the choices made, using a common sense approach, in rounding off. Only one significant figure is maintained in $P_{\bar{T}}$, and it was decided to report $\bar{T}$ to the corresponding number of decimal places.

Example 2.3

The ultimate strength, X, of an unknown alloy is to be determined by pulling five samples in a tensile test machine. The first four tests give values of 65,340, 68,188, 67,723, and 66,453 psi. Within what range (at 20 : 1 odds) would we anticipate the fifth value to lie?

The range at $20:1$ odds (or 95% confidence) for the next single reading would be given by

$$\overline{X} \pm P_X$$

Using Eqs. (2.11) and (2.12), we calculate

$$\overline{X} = 66,926 \text{ psi}$$
$$S_X = 1,287 \text{ psi}$$

Since $N = 4$, from Table A.2 we find $t = 3.182$ and

$$P_X = tS_X = (3.182)(1,287) = 4,095 \text{ psi}$$

The 95% confidence range for the fifth reading would then be

$$\overline{X} \pm P_X = 66,926 \text{ psi} \pm 4,095 \text{ psi}$$

A common sense argument can be made for reporting this as $67,000 \pm 4,000$ psi.

The use of an "odds" statement (as in the previous example) rather than a "% confidence" statement is sometimes encountered when uncertainties are quoted. Although $19:1$ odds (19 chances out of 20) strictly corresponds to a fractional value of 0.95, in practice it is commonplace for $20:1$ odds and 95% confidence to be used interchangeably. The same holds true for $2:1$ odds and 68% confidence, although two chances out of three corresponds to a fractional value of 0.667.

2-3.3   Statistical Rejection of "Wild" Readings from a Sample Population

When a sample of N readings of a particular variable is examined, sometimes there are readings that appear to be significantly out of line with the other readings. Such points are often called "outliers" or "wild" points. If obvious, verifiable problems with the experiment when such points were taken are identified, then they can be discarded. However, the more common situation is when there is no obvious or verifiable reason for the large deviation of these data points. In this case, we must use a statistical criterion to identity points that might be considered for rejection. There is no other justifiable way to "throw away" data points.

One method that has achieved relatively wide acceptance is Chauvenet's criterion, and it defines an acceptable scatter (from a probability viewpoint) around the mean value from a given sample of N readings from the same parent population. The criterion specifies that all points should be retained

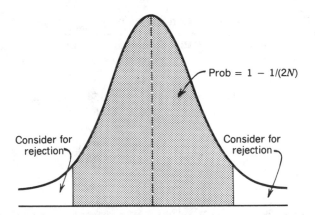

Figure 2.10 Graphic representation of Chauvenet's criterion.

that fall within a band around the mean value that corresponds to a probability of $1 - 1/(2N)$. Stated differently, we can say that data points can be considered for rejection if the probability of obtaining their deviation from the mean value is less than $1/(2N)$. This criterion is shown schematically in Figure 2.10.

As an illustration of this criterion, consider a test in which six readings are taken at a "constant" test condition. According to Chauvenet's criterion, all of the points that fall within a probability band around the mean of $1 - (1/12)$ or 0.917 must be retained. This probability can be related to a definite deviation away from the mean value by using the Gaussian probabilities in Table A.1. (Note that Chauvenet's criterion does not use t-distribution probabilities, even when N is small.) For a probability of 0.917, the nondimensional deviation τ equals 1.73 on interpolating in the table. Thus,

$$\tau = \frac{X_i - \overline{X}}{S_X} = \frac{x_{max}}{S_X} \tag{2.22}$$

where x_{max} is the maximum deviation allowed away from the mean value $\overline{X}$ of the six readings and S_X is the precision index (standard deviation) of the sample of six points. Therefore, all readings that deviate from the mean by more than $(1.73)S_X$ can be rejected. Then a new mean value and a new precision index are calculated from the readings that remain. No further application of the criterion to the sample is allowed—Chauvenet's criterion can be applied only once for a given sample of readings.

This statistical method for identifying readings that might be considered for rejection depends on the number of readings, N, in the sample. Table 2.1 gives the maximum acceptable deviations for various sample sizes. Values for other Ns can be determined using the Gaussian probabilities in Table A.1.

TABLE 2.1 Chauvenet's Criterion for Rejecting a Reading

Number of Readings (N)	Ratio of Maximum Acceptable Deviation to Precision Index (x_{max}/S_X)
3	1.38
4	1.54
5	1.65
6	1.73
7	1.80
8	1.87
9	1.91
10	1.96
15	2.13
20	2.24
25	2.33
50	2.57
100	2.81
300	3.14
500	3.29
1,000	3.48

Example 2.4

A laboratory has determined the ultimate strength of five samples of an alloy on their tensile testing machine, and the values (in psi) are reported as

i	X_i	$\lvert(X_i - \bar{X})/S_X\rvert$
1	65,300	0.26
2	68,000	0.52
3	67,700	0.49
4	43,600	1.78
5	67,900	0.51

Calculation of $\bar{X}$ and S_X gives

$$\bar{X} = 62,500 \text{ psi}$$

$$S_X = 10,624 \text{ psi}$$

Having noticed that point number 4 appears out of line with the others, we apply Chauvenet's criterion: for

$$N = 5, \qquad x_{max}/S_X = 1.65$$

We see from the third column that point number 4 is indeed outside the range defined as statistically acceptable by the criterion.

If point 4 is rejected, then

$$\bar{X} = 67,225 \text{ psi}$$

$$S_X = 1,289 \text{ psi}$$

Comparing these with the values calculated previously, $\bar{X}$ is increased by only 8% whereas S_X is decreased by a factor of 8.

Example 2.5

A dynamic pressure reading (in inches of H_2O) was made by 10 different students from a pressure gauge connected to a Pitot probe exposed to "constant" flow conditions. The 10 readings were as follows:

i	X_i	$\|(X_i - \bar{X})/S_X\|$
1	1.15	0.07
2	1.14	0.00
3	1.01	0.87
4	1.10	0.27
5	1.11	0.20
6	1.09	0.33
7	1.10	0.27
8	1.10	0.27
9	1.55	2.73
10	1.09	0.33

Just looking down the X_i column, the ninth reading appears to be a "wild" point, but we have no basis on which to reject it except by statistical analysis. We apply Chauvenet's criterion as follows:

$$\bar{X} = 1.14$$

$$S_X = 0.15$$

for $N = 10$, $x_{max}/S_X = 1.96$ from Table 2.1. If Reading 9 is rejected, recalculation of $\bar{X}$ and S_X yields

$$\bar{X} = 1.10$$

$$S_X = 0.04$$

Note that rejection of the probable "wild" point changed the mean value by only about 4%, but S_X, the estimate of the standard deviation, is only about one-fourth of the value originally calculated.

2-3.4   Effect of Non-Gaussian Error Distributions

In our discussions up to this point, we have been assuming that the distributions of errors in measurements are normal, or Gaussian. Since it is a relatively rare instance in which a large enough number of measurements is obtained so that the probability distributions (and particularly the tails of the distributions) can be defined, we should consider two questions. First, when is it inadvisable to assume normality? Second, what is the order-of-magnitude of the effect if the errors are from a nonnormal distribution but normality is assumed?

In answer to the first question, it is inadvisable to assume normality when low-probability events (extreme values of the variables) are of interest. In these cases, the details of the tails of the error distributions are important and the assumption of normality can lead to results and conclusions that are seriously in error. Tests modeling failures and/or accidents at nuclear power plants, the effects of natural disasters (hundred-year floods, etc.), and similar cases depend on the tails of the distributions, and erroneous results can have a significant effect on human life and comfort. Such cases should not be investigated using assumptions of normality and are outside the scope of this book.

The second question is considered for cases in which low-probability events are not of interest and the assumption of normality is made. Using the methods to be discussed in the next chapter, Kline and McClintock [3] investigated the propagation of precision limits for a case in which the result was equal to the sum of two measured variables. They considered error distributions that were normal, sinusoidal (one-half period), and triangular. Their results showed that propagation through the uncertainty analysis expression (for desired odds of 19:1) produced intervals corresponding to 19:1 odds for the normal distribution, 16.6:1 odds for the sinusoidal distribution, and 17.2:1 odds for the triangular distribution. These odds correspond to 95.0, 94.3, and 94.5% confidence, respectively. These results come into even sharper perspective when we note that using 2.00 rather than 1.96 for t when N is large gives a (tS) corresponding to 95.5% confidence rather than 95.0%. Investigations by other researchers assuming other nonnormal distributions have led to results similar to those of Kline and McClintock.

We can therefore conclude that *the estimates of the precision limits P and also the uncertainty in the result that is determined from the uncertainty analysis procedures outlined in later chapters are relatively insensitive to deviations from normality in the error distributions of the measured variables.* We can thus feel comfortable with the assumption of normality except in situations in which the details of the tails of the distributions are important.

2-4   BIAS ERROR ESTIMATION

Consider the situation shown in Figure 2.11, where the distribution of a sample of readings of a variable X is shown. In Chapter 1 we discussed the fact that we want to give a specification of the "best" value of X and some estimate of the uncertainty associated with the measurement. The uncertainty is made up of two components—bias and precision.

In the preceding sections of this chapter, we have seen that a quantification of the uncertainty due to the precision errors in a measurement can be achieved by using tS, where S is the precision index (standard deviation) of the sample of N readings and t is the value from the t-distribution that gives a particular confidence level, which we will generally take to be 95%. We thus have the estimation of this component of the uncertainty on a reasonably firm statistical foundation.

But what about the bias contribution to the uncertainty? There is no equivalent of the precision index that can be calculated from a sample of N readings and that can be used for a quantitative description of the bias. The bias is the same in every reading, and we will never know what it is unless we know the true value of X. Unfortunately, we do not know the true value of X in real experiments. This leaves us in somewhat of a quandary—how can we quantify the bias β and what equivalent of a statistical confidence level can we use?

The approach used in the Standard [1] is to define a bias limit, B, that is an estimated value of the limit of a confidence interval on the true value of the

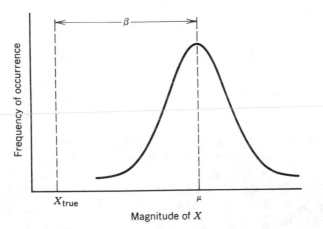

Figure 2.11   Bias and precision errors in a sample of readings of a variable X.

bias, β. As we have mentioned previously, an estimate of the bias as being between plus and minus infinity will certainly be true, but it will not be very useful to anyone. The bias limit is estimated at a level of surety corresponding to the confidence level with which tS is determined. Thus, when the precision limit $P = tS$ is determined at the specified 95% confidence level, the bias limit B is also estimated at 95% confidence. This can be interpreted to mean that we are 95% sure that the magnitude of the bias β is equal to or less than B.

When we discuss or report the value of a measured quantity, we should also include the estimates of the bias and precision limits and the associated confidence level.

In later chapters, we will take a detailed look at how various biases and precision errors in measured variables combine into an overall uncertainty in an experimental result, and we will also consider ways in which actual estimates of the bias and precision components can be made.

2-5   OVERALL UNCERTAINTY IN A MEASUREMENT

In the preceding sections, we have introduced the concepts of precision limit P and bias limit B as estimates of the limits of the 95% confidence intervals for the precision errors and bias errors in a measured variable. Once these estimates are obtained, it is reasonable to consider how P and B should be combined to obtain the overall uncertainty U in a measured variable.

Two methods of combination are presented in the Standard [1]. The first method is combination by the root-sum-square (RSS):

$$U_X = \left(B^2 + P_X^2 \right)^{1/2} \tag{2.23}$$

where

$$P_X = tS_X \tag{2.18}$$

or

$$U_{\bar{X}} = \left[B^2 + \left(P_{\bar{X}} \right)^2 \right]^{1/2} \tag{2.24}$$

where

$$P_{\bar{X}} = tS_{\bar{X}} = tS_X/\sqrt{N} \tag{2.19}$$

To be consistent with the nomenclature of the Standard, U_{RSS} is said to be a 95% coverage estimate when B and P are 95% confidence values. The interval $\pm U_X$ about X_i or $\pm U_{\bar{X}}$ about $\bar{X}$ will cover the true value of X about 95 times out of 100 (or at about 20 : 1 odds).

The second method presented in the Standard is combination by straight addition (ADD):

$$U_{X_{\text{ADD}}} = B + P_X \tag{2.25}$$

or

$$U_{\bar{X}_{ADD}} = B + P_{\bar{X}} \tag{2.26}$$

According to the Standard, when B and P are 95% confidence estimates, U_{ADD} provides approximately 99% coverage when neither B nor P is negligible relative to the other. However, as is obvious from Eqs. (2.25) and (2.26), U_{ADD} provides about 95% coverage when either B or P is negligible relative to the other.

The authors prefer the RSS combination method, and it will be assumed throughout the book unless otherwise indicated.

REFERENCES

1. *Measurement Uncertainty*, ANSI/ASME PTC 19.1-1985 Part 1, 1986.
2. Bowker, A. H., and Lieberman, G. J., *Engineering Statistics*, Prentice-Hall, Englewood Cliffs, NJ, 1959.
3. Kline, S. J., and McClintock, F. A., "Describing Uncertainties in Single-Sample Experiments," *Mechanical Engineering*, Vol 75, Jan. 1953, pp 3–8.

CHAPTER 2

Problems

2.1 A Gaussian distribution has a mean, μ, of 6.0 and a standard deviation, σ, of 2.0. Find the probability of a single reading from this distribution being (a) between 5.0 and 7.0, (b) between 7.0 and 9.0, (c) between 5.4 and 7.8, and (d) equal to or less than 6.4.

2.2 An electric motor manufacturer has found after many measurements that the mean value of the efficiency of his motors is 0.94. Variations in the motors and random errors in the test equipment and test procedure cause a standard deviation of 0.05 in the efficiency determination. With this same test arrangement, what is the probability that a motor's efficiency may be determined to be 1.00 or greater?

2.3 The standard deviation for using a certain radar speed detector is 0.6 mph for vehicle speeds around 35 mph. If the police use this detector, what is the probability that you would be clocked at below the speed limit if you were going 36 mph in a 35 mph speed zone?

2.4 A military cargo parachute has an automatic opening device set to switch at 200 m above the ground. The manufacturer of the device has specified that the standard deviation is 80 m about the nominal altitude setting. If the parachute must open 50 m above the ground to prevent damage to the equipment, what percentage of the loads will be damaged?

2.5   An accurate ohmmeter is used to check the resistance of a large number of 2 Ω resistors. Fifty percent of the readings fall between 1.8 Ω and 2.2 Ω. What is an estimate of the standard deviation for the resistance value of the resistors and within what range would the resistance of 95% of the resistors fall?

2.6   A pressure measurement system is known to have a standard deviation of 2 psi. If a single pressure reading of 60 psi is read, within what range (at 95% confidence) would we expect the mean, μ, to fall for a large number of readings at this steady condition?

2.7   The value of a variable is measured nine times, and the precision limit of the mean is calculated to be 10%. How many measurements would you estimate would be necessary to find the mean with a precision limit of 5%?

2.8   The precision index, S, for a sample of four readings has been calculated to be 1.00. What are the values of the precision limit of the readings and of the precision limit of the mean of the readings?

2.9   A thermocouple is placed in a constant temperature medium. Successive readings of 108.5°F, 107.6°F, 109.2°F, 108.3°F, 109.0°F, 108.2°F, and 107.8°F are made. Within what range would you estimate (at 20-1 odds) the next reading will fall? Within what 95% confidence range would you estimate the parent population mean (μ) to lie?

2.10   The ACME Co. produces Model 12B Roadrunner Inhibitors that are supposed to have a length of 2.00 m. You have been hired to determine the minimum number of samples (N) necessary for averaging if the plant manager must certify that the sample mean length ($\overline{X}$) is within 0.5% of the parent population mean (μ) with 95% confidence. You have been permitted to take length readings from nine randomly-chosen Model 12B's, and the results (in m) were 1.96, 1.95, 1.97, 1.94, 1.98, 1.95, 1.97, 1.99, and 1.96. Estimate N and state all of your assumptions.

2.11   The spring constant of a certain spring has been determined multiple times (in lb/in) as 10.36, 10.43, 10.41, 10.48, 10.39, 10.46, and 10.42. What is the best estimate (with 95% confidence) of the range which contains the parent population mean of the spring constant?

2.12   How would the range in Problem 2.11 vary if the confidence level were changed to (a) 90%, (b) 99%, (c) 99.5%, and (d) 99.9%?

2.13   If 30 data readings are taken, according to Chauvenet's criterion what is the ratio of the maximum acceptable deviation to the precision index?

2.14   Ten different people measured the diameter of a circular plate and obtained (in cm) 6.57, 6.60, 6.51, 6.61, 6.59, 6.65, 6.61, 6.63, 6.62, and

6.64. Compare the sample mean and precision index before and after application of Chauvenet's criterion.

2.15 Twelve values of the calibration constant of an oxygen bomb calorimeter were determined (in cal/°F): 1385, 1381, 1376, 1393, 1387, 1400, 1391, 1384, 1394, 1387, 1343, and 1382. Using these data, determine the best estimate of the mean calibration constant with 95% confidence.

2.16 In a metal casting facility, castings of specific types are weighed at random to check for consistency. The precision limit associated with the weight of a specific type casting (considering material variation and instrument readability) is ±10 lb and the bias limit associated with the scale is ±10 lb (both at 95% confidence). If the average weight of a casting is 500 lb, within what range would any one weight measurement fall with 95% confidence? Within what range does the "true" average weight fall? If the scale were replaced with an identical model, within what range would we expect a single weight measurement to fall?

3

PLANNING AN EXPERIMENT: GENERAL UNCERTAINTY ANALYSIS

In the previous chapter, we discussed the idea of uncertainty in the measured value of a single variable. In many cases, however, we do not directly measure the value of the experimental result. Instead, we measure the values of several variables and combine these in a data reduction equation to obtain the value of the desired result. For example, suppose that we want to answer the question "What is the density of the air in a pressurized tank?" Not having a "density meter" at hand, we consider what physical principles are available that will yield the value of the density if we measure the values of some other variables. If conditions are such that the equation of state of an ideal gas applies, then

$$p = \rho RT \tag{3.1}$$

offers some hope. If we find the gas constant for air and are able to measure the pressure and temperature, we can combine those values using Eq. (3.1) as the data reduction equation and obtain a value for ρ.

In some experiments, we may have several data reduction equations that each provides a result, with the purpose of the experiment being to determine how the results depend on one another. Such a case would occur if we wanted to describe the behavior of the heat transfer from a cubical object immersed in a fluid stream. We know that such behavior is usually described for a particular geometry using the nondimensional groups Nusselt number Nu, Reynolds number Re, and Prandtl number Pr. Each of these is a combination of variables that must be measured or found from tabulated data (as in the case of fluid properties), and each therefore constitutes a data reduction equation. The experiment would be designed and executed to determine the relationship between the three nondimensional groups (the results from the three data reduction equations) and perhaps to produce a correlation equation relating Nu, Re, and Pr.

3-1 PROPAGATION OF UNCERTAINTIES

What uncertainty should be associated with experimental results determined using a data reduction equation? The measurements of the variables (temperature, pressure, velocity, etc.) have uncertainties associated with them, and the values of the material properties that we obtain from reference sources also have uncertainties. *How do the uncertainties in the individual variables propagate through a data reduction equation into a result?* This is a key question in experimentation, and its answer is found using uncertainty analysis.

In the planning phase of an experimental program, the approach we use considers only the general, or overall, measurement uncertainties and not the details of the bias and precision components. This approach will be termed *general uncertainty analysis*. It makes sense to consider only the overall uncertainty in each measured variable at this stage rather than worry about which part of the uncertainty will be due to bias and which part will be due to precision errors. In what we are calling the planning phase, we are trying to identify which experiment or experiments will give us a chance to answer the question we have. In general, the particular equipment and instruments will not have been chosen—whether a temperature, for instance, will be measured with a thermocouple, thermometer, thermistor, RTD, or optical pyrometer is yet to be decided.

In a sense, at this stage all contributors to the uncertainty can be considered to be random. Since no particular transducer or instrument has been chosen from the population of all transducers and instruments available, any bias is just as likely to be positive as negative. Thus, we can view the general uncertainty analysis performed in the planning phase as a special case in which for each measured variable X_i, $B_{X_i} = 0$ and therefore $U_{X_i} = P_{X_i}$. Of course, at this stage there are usually no samples from which to compute statistical estimates of P_{X_i} using the methods of Chapter 2. Often, parametric studies are made for an assumed range of uncertainties in the variables. This will be demonstrated later in this chapter.

Once past the planning phase in an experimental program, it is desirable and useful to consider separately the propagation of the bias and the precision errors into the result. This is consistent with the method prescribed in the Standard [1], and it will be termed *detailed uncertainty analysis* as the details of the bias and precision components of the uncertainties are considered. The techniques of detailed uncertainty analysis are presented in Chapter 4.

In the following sections, we will describe the technique of general uncertainty analysis and some rules and helpful hints for its application, discuss the factors that must be considered in applying uncertainty analysis in the planning phase of an experimental program, and present comprehensive examples of the application of general uncertainty analysis in planning an experiment.

3-2 GENERAL UNCERTAINTY ANALYSIS

Consider a general case in which an experimental result, r, is a function of J variables X_i

$$r = r(X_1, X_2, \ldots, X_J) \tag{3.2}$$

Equation (3.2) is the data reduction equation used for determining r from the measured values of the variables X_i. Then the uncertainty in the result is given by

$$U_r = \left[\left(\frac{\partial r}{\partial X_1} U_{X_1} \right)^2 + \left(\frac{\partial r}{\partial X_2} U_{X_2} \right)^2 + \cdots + \left(\frac{\partial r}{\partial X_J} U_{X_J} \right)^2 \right]^{1/2} \tag{3.3}$$

where the U_{X_i} are the uncertainties in the measured variables X_i.

The development of Eq. (3.3) is presented and discussed in Appendix B. It is assumed that the relationship given by (3.2) is continuous and has continuous derivatives in the domain of interest, that the measured variables X_i are independent of one another, and that the uncertainties in the measured variables are independent of one another.

If the partial derivatives are defined [1] as "absolute sensitivity coefficients" so that

$$\theta_i = \frac{\partial r}{\partial X_i} \tag{3.4}$$

then Eq. (3.3) can be written as

$$U_r = \left[\sum_{i=1}^{J} \theta_i^2 U_{X_i}^2 \right]^{1/2} \tag{3.5}$$

In (3.3) and (3.5), all of the uncertainties (U_{X_i}) should be expressed with the same odds or coverage. In most cases, 95% coverage (20–1 odds) is used with the uncertainty in the result then also being at 95% coverage. In the planning phase it is implicitly understood that the values that we assume for the measurement uncertainties are all at the same coverage.

3-3   APPLICATION OF GENERAL UNCERTAINTY ANALYSIS

3-3.1   A Simple Case

Consider the case we discussed earlier. The question we wish to answer is "What is the density of the air in the pressurized tank?" The ideal gas law

$$p = \rho RT \tag{3.6}$$

relates the desired experimental result ρ to two quantities (pressure and temperature) that are rather commonly measured and to a quantity (the gas constant for air) that is known with a great degree of certainty and is available in any number of reference books.

Before proceeding with the details of the design of the experiment, a general uncertainty analysis should be performed to investigate the response of the result to uncertainties in the variables that will be measured and/or obtained from reference sources. In this section, we will discuss several general rules and "helpful hints" that will help to avoid some of the most common mistakes made in applying uncertainty analysis.

RULE 1:   Always solve the data reduction equation for the experimental result before doing an uncertainty analysis.

Applying Rule 1 in the current case gives

$$\rho = \frac{p}{RT} \tag{3.7}$$

The general uncertainty analysis expression [Eq. (3.3)] becomes, for this particular case,

$$U_\rho^2 = \left(\frac{\partial \rho}{\partial p} U_p \right)^2 + \left(\frac{\partial \rho}{\partial R} U_R \right)^2 + \left(\frac{\partial \rho}{\partial T} U_T \right)^2 \tag{3.8}$$

The partial derivatives are

$$\frac{\partial \rho}{\partial p} = \frac{1}{RT} \tag{3.9}$$

$$\frac{\partial \rho}{\partial R} = -\frac{p}{R^2 T} \tag{3.10}$$

$$\frac{\partial \rho}{\partial T} = -\frac{p}{RT^2} \tag{3.11}$$

and substituting back into (3.8) we obtain

$$U_\rho^2 = \left(\frac{1}{RT}U_p\right)^2 + \left(\frac{-p}{R^2T}U_R\right)^2 + \left(\frac{-p}{RT^2}U_T\right)^2 \qquad (3.12)$$

The form of the equation obtained at this stage is often more algebraically complex than necessary. Simplification can generally be achieved by dividing the equation by the square of the experimental result. In this particular case, we note from Eq. (3.7) that dividing by ρ^2 and $(p/RT)^2$ is equivalent, and we divide the left-hand side of (3.12) by ρ^2 and the right-hand side of (3.12) by $(p/RT)^2$ to obtain

$$\left(\frac{U_\rho}{\rho}\right)^2 = \left[\left(\frac{RT}{p}\right)\frac{1}{RT}U_p\right]^2 + \left[\left(\frac{RT}{p}\right)\frac{-p}{R^2T}U_R\right]^2 + \left[\left(\frac{RT}{p}\right)\frac{-p}{RT^2}U_T\right]^2 \quad (3.13)$$

or

$$\frac{U_\rho^2}{\rho^2} = \frac{U_p^2}{p^2} + \frac{U_R^2}{R^2} + \frac{U_T^2}{T^2} \qquad (3.14)$$

which is a simpler form than (3.12).

RULE 2: **Always try dividing the uncertainty analysis expression by the experimental result to see if algebraic simplification can be achieved.**

The authors have observed that the chances of making an algebraic error are lessened by applying Rule 2 as follows. Divide the general expression (3.3) by the result, r, and recast it into the form

$$\left(\frac{U_r}{r}\right)^2 = \left(\frac{1}{r}\frac{\partial r}{\partial X_1}U_{X_1}\right)^2 + \left(\frac{1}{r}\frac{\partial r}{\partial X_2}U_{X_2}\right)^2 + \cdots + \left(\frac{1}{r}\frac{\partial r}{\partial X_J}U_{X_J}\right)^2 \quad (3.15)$$

When the partial derivatives are taken, divide them by the result (making use of the data reduction equation) and then substitute into Eq. (3.15).

If this had been done in the analysis above, Eqs. (3.9)–(3.11) would have been replaced by

$$\left(\frac{1}{\rho}\right)\frac{\partial\rho}{\partial p} = \left(\frac{RT}{p}\right)\frac{1}{RT} = \frac{1}{p} \qquad (3.16)$$

$$\left(\frac{1}{\rho}\right)\frac{\partial\rho}{\partial R} = \left(\frac{RT}{p}\right)\frac{-p}{R^2T} = -\frac{1}{R} \qquad (3.17)$$

$$\left(\frac{1}{\rho}\right)\frac{\partial\rho}{\partial T} = \left(\frac{RT}{p}\right)\frac{-p}{RT^2} = -\frac{1}{T} \qquad (3.18)$$

and Eq. (3.14) would have been obtained without the intermediate steps of Eqs. (3.12) and (3.13).

Equation (3.14) relates the uncertainty in the experimental result, ρ, to the uncertainties in the measured variables and the gas constant. The values of universal constants (gas constant, pi, Avogadro's number, etc.) are typically known with much greater accuracy than the measurements that are made in most experiments. It is common practice to assume the uncertainties in such quantities are negligible and to set them equal to zero. Assuming that U_R is zero, our uncertainty expression then becomes

$$\left(\frac{U_\rho}{\rho}\right)^2 = \left(\frac{U_p}{p}\right)^2 + \left(\frac{U_T}{T}\right)^2 \tag{3.19}$$

We are now in a position to answer many types of questions about the proposed experiment. Two questions that might be of interest are considered in the following examples.

Example 3.1

A pressurized air tank (Figure 3.1) is nominally at ambient temperature (25°C). How accurately can the density be determined if the temperature is measured with an uncertainty of 2°C and the tank pressure is measured with an uncertainty of 1%?

The data reduction equation (3.7) is

$$\rho = \frac{p}{RT}$$

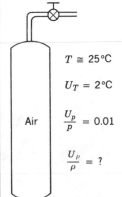

$T \cong 25\,°\mathrm{C}$

$U_T = 2\,°\mathrm{C}$

$\dfrac{U_p}{p} = 0.01$

$\dfrac{U_\rho}{\rho} = ?$

Figure 3.1   Sketch for Example 3.1.

and the general uncertainty analysis expression (3.19) is

$$\left(\frac{U_\rho}{\rho}\right)^2 = \left(\frac{U_p}{p}\right)^2 + \left(\frac{U_T}{T}\right)^2$$

Values for the variables are

$$U_T = 2°C(= 2 \text{ K})$$
$$T = 25 + 273 = 298 \text{ K}$$
$$(U_p/p) = 0.01$$

since the 1% uncertainty in the pressure measurement is interpreted mathematically as

$$U_p = 0.01p$$

Substitution of the numerical values into (3.19) yields

$$\left(\frac{U_\rho}{\rho}\right)^2 = (0.01)^2 + \left(\frac{2 \text{ K}}{298 \text{ K}}\right)^2$$

$$\left(\frac{U_\rho}{\rho}\right)^2 = 1.0 \times 10^{-4} + 0.45 \times 10^{-4} = 1.45 \times 10^{-4}$$

and taking the square root gives the uncertainty in the density

$$U_\rho/\rho = 0.012 \text{ or } 1.2\%$$

Example 3.2

For the physical situation in the previous example, suppose the density determination must be within 0.5% to be useful. If the temperature measurement can be made to within 1°C, how accurate must the pressure measurement be? (See Figure 3.2.)

Again Eq. (3.19) is used

$$\left(\frac{U_\rho}{\rho}\right)^2 = \left(\frac{U_p}{p}\right)^2 + \left(\frac{U_T}{T}\right)^2$$

with numerical values of

$$U_T = 1°C \, (= 1 \text{ K})$$
$$T = 25 + 273 = 298 \text{ K}$$
$$\frac{U_\rho}{\rho} = 0.005$$

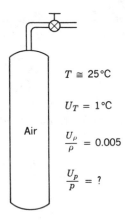

$T \cong 25\,°C$

$U_T = 1\,°C$

$\dfrac{U_\rho}{\rho} = 0.005$

$\dfrac{U_p}{p} = ?$

Figure 3.2   Sketch for Example 3.2.

so that we obtain

$$(0.005)^2 = \left(\frac{U_p}{p}\right)^2 + \left(\frac{1\,K}{298\,K}\right)^2$$

or

$$\left(\frac{U_p}{p}\right)^2 = 1.37 \times 10^{-5}$$

$$\frac{U_p}{p} = 0.0037 \text{ or } 0.37\%$$

The pressure measurement would have to be made with an uncertainty of less than 0.37% for the density measurement to meet the specifications.

3-3.2   A Special Functional Form

A very useful specific form of Eqs. (3.3) and (3.15) is obtained when the data reduction equation [Eq. (3.2)] has the form

$$r = kX_1^a X_2^b X_3^c \ldots \tag{3.20}$$

where the exponents may be positive or negative constants and k is a constant. Application of Eq. (3.15) to the relationship of (3.20) yields

$$\left(\frac{U_r}{r}\right)^2 = a^2\left(\frac{U_{X_1}}{X_1}\right)^2 + b^2\left(\frac{U_{X_2}}{X_2}\right)^2 + c^2\left(\frac{U_{X_3}}{X_3}\right)^2 + \cdots \tag{3.21}$$

A data reduction equation of the form (3.20) is thus especially easy to work with, as the result of the uncertainty analysis may be written down by inspection in the form (3.21) without any partial differentiation and subsequent algebraic manipulation. One must remember, however, that the X_is in (3.20) represent variables that are directly measured. Thus,

$$W = I^2R \tag{3.22}$$

is of the requisite form when the current I and the resistance R are measured, but

$$Y = Z \sin \theta \tag{3.23}$$

is not if Z and θ are the measured variables, and

$$V = Z(p_2 - p_1)^{1/2} \tag{3.24}$$

is not if p_2 and p_1 are measured separately. If the difference $(p_2 - p_1)$ is measured directly, however, then (3.24) is in a form where (3.21) can be applied.

It is interesting to note that the signs of the exponents in (3.20) have no effect on the propagation of the measurement uncertainties into the experimental result as all exponents appear squared in (3.21).

Inspection of Eq. (3.21) reveals that, in the propagation process, the influence of uncertainties in variables raised to a power with magnitude greater than one is magnified. Conversely, the influence of the uncertainties in variables raised to a power with magnitude less than one is diminished.

Example 3.3

We again plan to use the ideal gas law to determine the density of a gas. The density determination must be accurate to within 0.5%, and we can measure the temperature (nominally 25°C) to within 1°C. What is the maximum allowable uncertainty in the pressure measurement? (See Figure 3.3.)

The ideal gas law

$$p = \rho RT$$

is again our data reduction equation. This time we note that this is in the special form of Eq. (3.20), and assuming zero uncertainty in the gas constant we can write down by inspection

$$\left(\frac{U_p}{p}\right)^2 = \left(\frac{U_\rho}{\rho}\right)^2 + \left(\frac{U_T}{T}\right)^2$$

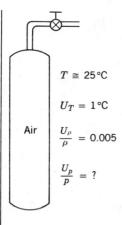

$T \cong 25\,°C$

$U_T = 1\,°C$

Air

$\dfrac{U_\rho}{\rho} = 0.005$

$\dfrac{U_p}{p} = ?$

Figure 3.3   Sketch for Example 3.3.

and substituting in the given information we can find

$$\left(\frac{U_p}{p}\right)^2 = (0.005)^2 + \left(\frac{1\,K}{298\,K}\right)^2$$

and the required uncertainty in the pressure measurement is

$$U_p/p = 0.0061 \text{ or } 0.61\%$$

BUT WAIT! THIS IS INCORRECT. We did not solve for the experimental result (ρ) before we did the uncertainty analysis. If we follow Rule 1 and solve for the result first, we have

$$\rho = p^1 R^{-1} T^{-1}$$

and using (3.21) and assuming U_R negligible

$$\left(\frac{U_\rho}{\rho}\right)^2 = \left(\frac{U_p}{p}\right)^2 + \left(\frac{U_T}{T}\right)^2$$

Substituting in the given information we find

$$(0.005)^2 = \left(\frac{U_p}{p}\right)^2 + \left(\frac{1\,K}{298\,K}\right)^2$$

so that

$$\left(\frac{U_p}{p}\right)^2 = (0.005)^2 - (0.0034)^2$$

and the required uncertainty in the pressure measurement is

$$\frac{U_p}{p} = 0.0037 \text{ or } 0.37\%$$

which agrees with the result found in Example 3.2. Note that the required uncertainty was erroneously found to be 1.6 times the correct value when the incorrect analysis was done.

This example is intended to show how easily a mistake can be made when the data reduction equation is of the special form and the uncertainty expression is written down by inspection. *It is imperative that the data reduction equation be solved for the experimental result before the uncertainty analysis is begun.*

Example 3.4

The power W drawn by a resistive load in a dc circuit (Figure 3.4) can be determined by measuring the voltage E and current I and using

$$W = EI$$

It can also be determined by measuring the resistance R and the current and using

$$W = I^2 R$$

If I, E, and R can all be measured with approximately equal uncertainties (on a percentage basis), which method of power determination is "best"?

All other things (cost, difficulty of measurement, etc.) being equal, the best method will be the one that produces a result with the least uncertainty. Both data reduction expressions are in the special form of Eq. (3.20). For

$$W = EI$$

$$\left(\frac{U_W}{W}\right)^2 = \left(\frac{U_E}{E}\right)^2 + \left(\frac{U_I}{I}\right)^2$$

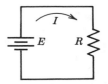

Figure 3.4   Sketch for Example 3.4.

and for

$$W = I^2 R$$

$$\left(\frac{U_W}{W}\right)^2 = (2)^2 \left(\frac{U_I}{I}\right)^2 + \left(\frac{U_R}{R}\right)^2$$

If the E, I, and R measurements have equal percentage uncertainties, then the second method produces a result with an uncertainty 1.58 times that in the result from the first method. This can be verified by substitution of any assumed percentage value of uncertainty for the measured variables.

This example is intended to emphasize the usefulness of uncertainty analysis in the planning phase of an experiment and the generality of the conclusions that can often be drawn. Note the conclusion we were able to make in this case without knowing the level of uncertainty in the measurements, what instruments might be available, or any other details.

Example 3.5

It is proposed that the shear modulus, M_S, be determined for an alloy by measuring the angular deformation θ produced when a torque T is applied to a cylindrical rod of the alloy with radius R and length L. The expression relating these variables [2] is

$$\theta = \frac{2LT}{\pi R^4 M_S}$$

We wish to examine the sensitivity of the experimental result to the uncertainties in the variables that must be measured before we proceed with a detailed experimental design.

The physical situation shown in Figure 3.5 (where torque T is given by

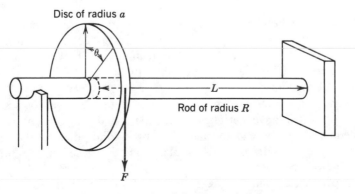

Figure 3.5   Sketch for Example 3.5.

aF) is described by the data reduction equation for the shear modulus

$$M_S = \frac{2LaF}{\pi R^4 \theta}$$

Noting that this is in the special form of Eq. (3.20) and assuming as usual that the factors 2 and π have zero uncertainties, general uncertainty analysis yields

$$\left(\frac{U_{M_S}}{M_S}\right)^2 = \left(\frac{U_L}{L}\right)^2 + \left(\frac{U_a}{a}\right)^2 + \left(\frac{U_F}{F}\right)^2 + 16\left(\frac{U_R}{R}\right)^2 + \left(\frac{U_\theta}{\theta}\right)^2$$

This shows the tremendous sensitivity of the uncertainty in the result to the uncertainty in variables whose exponents are large. In this case, if all variables are measured with 1% uncertainty, the uncertainty in M_S is 4.5%, whereas it still would be 4% if U_L, U_a, U_F, and U_θ were all assumed zero.

This example is intended to illustrate the utility of uncertainty analysis in identifying the measurements with which special care must be taken. Since this information can be determined in the planning phase of an experiment, it can be incorporated into the design of an experiment from the very beginning.

Note that the value assigned to the uncertainty in R should include the effects of the variation of radius along the length of the rod from the nominal value of R. The results of the general uncertainty analysis can thus guide the specification of the tolerances to which the rod must be manufactured for the experiment. If the uncertainty with which the shear modulus must be determined is known, then an estimate of the maximum permissible tolerances on the rod dimensions can be made.

3-4 USING UNCERTAINTY ANALYSIS IN PLANNING AN EXPERIMENT

In the previous sections of this chapter, we have discussed general uncertainty analysis and some initial examples of its application. It is worthwhile at this point to step back from the details and reemphasize how this technique fits into the experimental process and, particularly, how useful it is in the planning phase of experiments.

There is a question to which an experimental answer is sought. The planning phase of an experiment is the period during which the physical phenomena that might lead to such an answer—the potential data reduction equations—are considered. This is also sometimes called the preliminary design stage of an experiment.

It is important even at this stage that the allowable uncertainty in the result be known. This seems to be a point that bothers many, particularly those who are relative novices at experimentation. However, it is ridiculous to embark on an experimental effort without considering the "degree of goodness" necessary in the result. Here, we are talking about whether the result needs to be known within 0.1, 1, 10, or 50%, not whether 1.3 versus 1.4% is necessary. The degree of uncertainty allowable in the result can usually be estimated fairly well by considering the use that will be made of the result.

In the planning phase, general uncertainty analysis is used to investigate whether it is feasible to pursue a particular experiment, which measurements might be more important than others, which techniques might give better results than others, and similar matters. We are attempting to understand the behavior of the experimental result as a function of the uncertainties in the measured variables and to use this to our advantage, if possible. At this stage, we are not worrying about whether Brand A or Brand B instrument should be purchased or whether a temperature should be measured with a thermometer, thermocouple, or thermistor.

In using general uncertainty analysis, uncertainties must be assigned for the variables that will be measured and properties that typically will be found from reference sources. In years of teaching, the authors have observed an almost universal reluctance to estimate uncertainties. There seems to be a feeling that "out there" somewhere is an expert who knows what all these uncertainties *really* are and who will suddenly appear and scoff at any estimates we make. This is nonsense! Uncertainty analysis in the planning phase can be viewed as a spreadsheet that allows us to play "what if" before making decisions on the specifics of an experimental design.

If there is not an experience base for estimating the uncertainties, then a parametric analysis using a range of values can be made. There are no penalties at this stage for making ridiculous estimates—if we want to see the effect on a result assuming a temperature measurement uncertainty of 0.00001°F, we are free to do so.

The primary point to be made is that a general uncertainty analysis should be used in the very initial stages of an experimental program. The information and insight gained are far out of proportion to the small amount of time the analysis takes, and parametric analysis using a range of assumed values for the uncertainties is perfectly acceptable.

In the following sections of this chapter, comprehensive examples of the use of general uncertainty analysis in the planning phase of two experimental programs are presented. Time that the reader spends in working through these examples in detail will be well spent.

3-5   ANALYSIS OF A PROPOSED PARTICULATE MEASURING SYSTEM

3-5.1   The Problem

A manufacturer needs to determine the solid particulate concentration in the exhaust gases from one of its process units. Monitoring the concentration will allow immediate recognition of unexpected and undesired changes in the

operating condition of the unit. The measurement is also necessary to determine and monitor compliance with air pollution regulations. The measurement
needs to be accurate within about 5%, with a 10% or greater inaccuracy being
unacceptable.

A representative of an instrument company has submitted a presentation
proposing that the manufacturer purchase and install the Model LT1000 Laser
Transmissometer System to measure the projected area concentration of the
particulate matter in the stack exhaust. He states that the system will meet the
5% uncertainty requirement.

Management has directed the engineering department to evaluate the
salesman's proposal and recommend whether the system should be purchased.

3-5.2   The Proposed Measurement Technique and System

A schematic of the laser transmissometer system is shown in Figure 3.6. A
laser beam passes through the exhaust gas stream and impinges on a transducer that measures the beam intensity. The physical process is described [3]

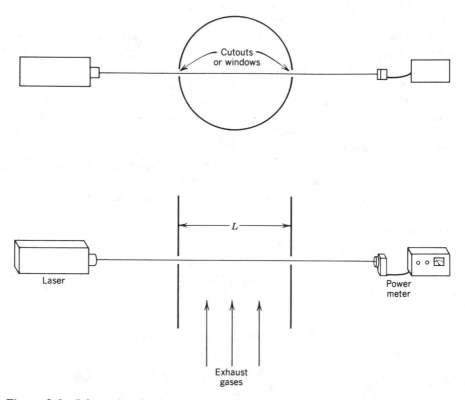

Figure 3.6   Schematic of a laser transmissometer system for monitoring particulate
concentrations in exhaust gases.

by the expression

$$T = \frac{I}{I_0} = e^{-CEL} \tag{3.25}$$

In this expression, I_0 is the intensity of the light beam exiting the laser, I is the transmitted intensity after the beam has passed through the scattering and absorbing medium of thickness L, and T is the fractional transmitted intensity (the transmittance). The projected area concentration C is the projected area of the particulates per unit volume of the medium.

The extinction coefficient E is a function of the wavelength of the laser light beam, the distribution of sizes and shapes of the particles, and the indices of refraction of the particles. E approaches an asymptotic value of 2.0 for certain optical conditions. The instrument company states in their proposal that based on their experience with exhaust stream measurements in similar plants, they expect the value of E will "be within a couple of percent of 2.0" over the range of operating conditions expected.

The system is adjusted so that when the laser beam travels through the stack with no exhaust flow, the power meter output reads 1.000. When there is flow through the stack, therefore, the output of the power meter corresponds directly to the value of the transmittance T. The instrument manufacturer specifies that the power meter is accurate to 1% of reading, or better.

3-5.3   Analysis of the Proposed Experiment

We have the assignment of deciding whether the proposed experimental approach will answer the question of interest (What is the projected area concentration of the particulates in the exhaust?) acceptably (with an uncertainty of about 5% of less). Although we may never have touched a laser or seen the particular exhaust stack in question, general uncertainty analysis gives us a logical, proven technique for evaluating the proposed experiment in the planning stage.

The data reduction equation is

$$T = e^{-CEL} \tag{3.26}$$

However, the experimental result is C, the particulate projected area concentration. Taking the natural logarithm of Eq. (3.26) and solving for C yields

$$C = -\frac{\ln T}{EL} \tag{3.27}$$

Since T, and not $\ln(T)$, is measured, Eq. (3.27) is not in the special product form, and the proper uncertainty analysis expression to use is Eq. (3.15). For this particular case

$$\left(\frac{U_C}{C}\right)^2 = \left[\left(\frac{1}{C}\right)\frac{\partial C}{\partial T}U_T\right]^2 + \left[\left(\frac{1}{C}\right)\frac{\partial C}{\partial E}U_E\right]^2 + \left[\left(\frac{1}{C}\right)\frac{\partial C}{\partial L}U_L\right]^2 \quad (3.28)$$

Proceeding with the partial differentiation gives

$$\left(\frac{1}{C}\right)\frac{\partial C}{\partial T} = \left\{\frac{-EL}{\ln T}\right\}\left(-\frac{1}{EL}\right)\left(\frac{1}{T}\right) = \frac{1}{\ln T}\frac{1}{T} \quad (3.29)$$

$$\left(\frac{1}{C}\right)\frac{\partial C}{\partial E} = \left\{\frac{-EL}{\ln T}\right\}\left(-\frac{\ln T}{L}\right)\left(\frac{1}{E^2}\right) = -\frac{1}{E} \quad (3.30)$$

$$\left(\frac{1}{C}\right)\frac{\partial C}{\partial L} = \left\{\frac{-EL}{\ln T}\right\}\left(-\frac{\ln T}{E}\right)\left(\frac{1}{L^2}\right) = -\frac{1}{L} \quad (3.31)$$

and substitution into (3.28) yields

$$\left(\frac{U_C}{C}\right)^2 = \left(\frac{1}{\ln T}\right)^2\left(\frac{U_T}{T}\right)^2 + \left(\frac{U_E}{E}\right)^2 + \left(\frac{U_L}{L}\right)^2 \quad (3.32)$$

This is the desired expression relating the uncertainty in the result to the uncertainties in the measured quantities T and L and the assumed value of E.

Equation (3.32) shows that the uncertainty in C depends not only on the uncertainties in L, T, and E, but also on the value of the transmittance itself. The uncertainty in C will thus vary with operating conditions even if the uncertainties in L, T, and E are all constant. In situations such as this, a parametric study is indicated.

In considering the physical situation, we can see that the transmittance T will vary between the limiting values 0 and 1. T will equal 1.0 when $I = I_0$, that is, when there are no particles in the flow through the stack. In the other limit, T will equal 0 when the flow is opaque and no light makes it through the flow. If we investigate the behavior of the uncertainty in the result for values of T between 0 and 1, the entire possible range of operating conditions will have been covered. In order to calculate values of U_C/C, we must first estimate values of the uncertainties in T, E, and L.

This is the point at which those unfamiliar with uncertainty analysis and inexperienced in either experimentation or the particular technique being considered typically feel uneasy. A reluctance to estimate uncertainty values seems to be a part of human nature. Such reluctance can be overcome by choosing a range of uncertainties that will almost certainly bracket the "true" uncertainties. In this case, we could calculate U_C/C for a set of T values between 0 and 1 for assumed uncertainties in T, E, and L of 0.1, 1, 10, and 50% if we wished.

In situations such as this, however, it is often quite revealing to initially choose all the uncertainties the same and proceed with the parametric study. If

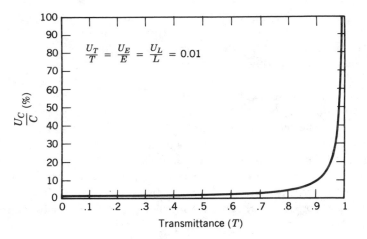

Figure 3.7   Uncertainty in the experimental result as a function of the transmittance.

we choose all uncertainties as 1%, then

$$\left(\frac{U_C}{C}\right)^2 = \left(\frac{1}{\ln T}\right)^2 (0.01)^2 + (0.01)^2 + (0.01)^2$$

$$= \left(\frac{1}{\ln T}\right)^2 (10^{-4}) + 2.0 \times 10^{-4} \qquad (3.33)$$

We note that as

$$T \to 0, \qquad \ln T \to -\infty$$

and the first term on the right-hand side approaches 0. Conversely, as

$$T \to 1, \qquad \ln T \to 0$$

and the first term on the right-hand side grows without bound. The behavior of Eq. (3.33) over the entire range of T from 0 to 1 is shown in Figure 3.7.

3-5.4   Implications of the Uncertainty Analysis Results

The results of the uncertainty analysis reveal a behavior that is certainly not obvious simply by looking at the expressions [Eqs. (3.26) and (3.27)] that model this physical phenomenon. Even if all measurements are made with 1% uncertainty, the uncertainty in the experimental result can be 10%, 20%, 100%, or greater.

It should be noted that this behavior does not occur at operating conditions that are unlikely to be encountered. To the contrary, this behavior is observed in the range of operating conditions one would anticipate being of interest. For there to be 30%, 40%, or more extinction of the light beam over a relatively short distance such as a typical stack diameter, the solid particle concentration in the exhaust gases would have to be tremendous. It would seem likely, therefore, that measurements with T less than 0.7 or 0.6 would occur at conditions unacceptable from a pollution standpoint.

We can conclude, then, that the technique as proposed is not acceptable. It could not be expected to yield results with uncertainty of 5–10% or less over the range of operating conditions one would expect to encounter.

It should be noted that this conclusion was reached without our having to investigate the details of the dependence of E, the extinction coefficient, on the nature of the particles. The results of the uncertainty analysis showed unacceptable behavior of the uncertainty in the experimental result even with an optimistic assumption of 1% uncertainty in the value of E.

3-5.5   Design Changes Indicated by the Uncertainty Analysis

In looking at the uncertainty analysis results plotted in Figure 3.7, we see that there would be a chance that the technique would be acceptable if the transmittance were always 0.8 or less. Since we are doing an uncertainty analysis in the planning phase of an experiment, we are free to play "what if" with almost no restrictions. In fact, playing "what if" at this stage of an experimental design should be encouraged.

Consider the expression for the transmittance

$$T = e^{-CEL} \tag{3.26}$$

For a given set of operating conditions, the characteristics of the exhaust flow are fixed and therefore C and E are fixed. The only way to cause a decrease in T, then, is to increase L, the path length of the light beam through the exhaust stream. Rather than recommend that a new, larger diameter stack be constructed(!), we might recommend that two mirrors on adjustable mounts be purchased and used as shown in Figure 3.8.

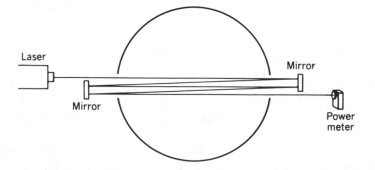

Figure 3.8   Multiple pass beam arrangement for the transmissometer.

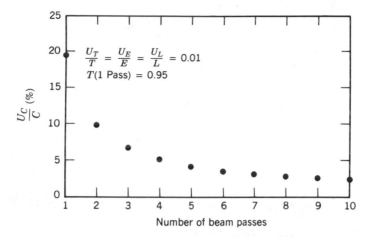

Figure 3.9 Uncertainty in the experimental result as a function of the number of beam passes.

The two additional mirrors allow multiple passes of the beam through the stack, thus increasing the path length L, decreasing the overall transmittance, and decreasing the uncertainty in the measurement of C. For example, consider the case in which all measurements are made with 1% uncertainty and the transmittance for one pass of the beam through the stack is 0.95. The effect of additional passes on the transmittance can be calculated from $T(n \text{ passes}) = (0.95)^n$. The behavior of the uncertainty in the result is shown in Figure 3.9 as a function of the number of beam passes through the stack.

The results of the above analysis indicate that the proposed system, with modifications, might be able to meet the requirements. Additional factors, such as the effect of laser output (I_0) variation with time and the probable behavior of E, should be investigated [4, 5] if the implementation of the technique is to be considered further.

3-6 ANALYSIS OF A PROPOSED HEAT TRANSFER EXPERIMENT

In this section, we use general uncertainty analysis to investigate the suitability of two techniques that might be used to determine the convective heat transfer coefficient for a circular cylinder immersed in an air flow. An extensive analysis and discussion of this case has been presented by Moffat [6].

3-6.1 The Problem

A client of our research laboratory needs to determine the behavior of the convective heat transfer from a circular cylinder of finite length that will be immersed in an air flow whose velocity might range from 1 to 100 m/sec. The physical situation and nomenclature are shown in Figure 3.10.

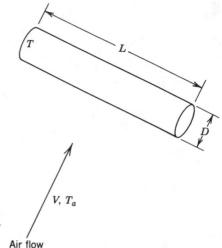

Figure 3.10 Finite length circular cylinder in crossflow.

The convective heat transfer coefficient h is defined by the relationship

$$q = hA(T - T_a) \qquad (3.34)$$

where q is the rate of convective heat transfer from the cylinder to the air stream, T is the temperature of the surface of the cylinder, T_a is the temperature of the oncoming air, and A is the surface area, given in this case by

$$A = \pi DL \qquad (3.35)$$

For T, T_a, and A remaining constant, h (and hence q) increases with increasing air velocity V.

The cylindrical tube to be tested is 0.152 m (6.0 in.) long with a diameter of 0.0254 m (1.0 in.) and a wall thickness of 1.27 mm (0.050 in.). The primary case of interest is for an air temperature of about 25°C with the cylinder about 20°C above that. The cylinder is made of aluminum, for which the specific heat has been found from a table [7] as 0.903 kJ/kg-C at 300 K. The mass of the tube has been measured as 0.020 kg.

The client wants to know h within 5% and must be assured that h can be determined within 10% before funding for the experiment will be allocated.

3-6.2 The Two Proposed Experimental Techniques

Two techniques have been suggested for determining the heat transfer coefficient. The first is a steady-state technique, and the second is a transient technique.

The Steady-State Technique. As shown schematically in Figure 3.10, the cylinder is mounted in a 25°C air stream of velocity V. Sufficient electric power is supplied to a resistance heater in the cylindrical tube so that the surface temperature reaches a steady-state value of about 45°C. We now apply the first law of thermodynamics to a control volume whose surface lies infinitesimally outside the surface of the cylinder. Assuming the radiative and conductive heat transfer losses to be negligible for purposes of this initial analysis, the energy balance yields

$$\text{Power in} = \text{Rate of convected energy out} \tag{3.36}$$

or

$$W = h(\pi DL)(T - T_a) \tag{3.37}$$

Solving for the experimental result,

$$h = \frac{W}{\pi DL(T - T_a)} \tag{3.38}$$

For a given air velocity, then, the convective heat transfer coefficient can be determined by measuring the electrical power W, the cylinder diameter D and length L, and the temperatures of the cylinder surface and the air stream.

The Transient Technique. This approach uses the characteristics of the response of a first-order system to a step change in the input to determine the heat transfer coefficient. The experimental setup is similar to that for the steady-state case. The cylinder is heated until it reaches a temperature T_1, say, and then the power is switched off. The cylinder is then cooled by the air stream, and its temperature T will eventually reach T_a. The temperature versus time behavior is shown in Figure 3.11.

The time taken for the cylinder temperature T to change 63.2% of the way from T_1 to T_a is equal to the time constant τ of the system. τ is related to the convective heat transfer coefficient by

$$\tau = \frac{Mc}{h(\pi DL)} \tag{3.39}$$

where M is the mass of the cylinder and c is the specific heat of the cylinder. Solving for the experimental result

$$h = \frac{Mc}{\pi DL\tau} \tag{3.40}$$

For a given air velocity, the heat transfer coefficient can be determined by

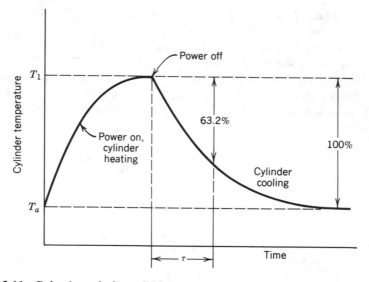

Figure 3.11 Behavior of the cylinder temperature in a test using the transient technique.

measuring the mass, diameter, and length of the cylinder, finding the specific heat of the cylinder, and measuring a time corresponding to τ.

In the above, radiative and conductive heat transfer losses are assumed negligible, h and c are assumed constant during the cooling process, and the cylinder is assumed to have negligible internal resistance (each point in the cylinder is at the same temperature T as every other point in the cylinder). This last assumption holds true for Biot numbers less than one-tenth [7].

To determine whether the techniques can produce results with acceptable uncertainty, if one technique has advantages over the other, or if there is any behavior in the uncertainty in the result that is not obvious from Eqs. (3.38) and (3.40), a general uncertainty analysis should be performed for each of the proposed techniques.

3-6.3   General Uncertainty Analysis: Steady-State Technique

The data reduction expression for the steady-state technique is given by Eq. (3.38)

$$ h = \frac{W}{\pi DL (T - T_{\mathrm{a}})} \tag{3.38} $$

If the temperatures are measured separately, this expression is not in the special product form [Eq. (3.20)] and the general uncertainty analysis must be done using Eq. (3.15). Leaving the details as an exercise for the reader, the analysis yields

$$\left(\frac{U_h}{h}\right)^2 = \left(\frac{U_W}{W}\right)^2 + \left(\frac{U_D}{D}\right)^2 + \left(\frac{U_L}{L}\right)^2 + \left(\frac{U_T}{(T - T_a)}\right)^2 + \left(\frac{U_{T_a}}{(T - T_a)}\right)^2$$

$$(3.41)$$

For the present analysis, assume that the temperature difference $(T - T_a)$ will be measured directly, perhaps with a differential thermocouple circuit. In this case, Eq. (3.38) is in the special product form, and we can write down by inspection

$$\left(\frac{U_h}{h}\right)^2 = \left(\frac{U_W}{W}\right)^2 + \left(\frac{U_D}{D}\right)^2 + \left(\frac{U_L}{L}\right)^2 + \left(\frac{U_{\Delta T}}{\Delta T}\right)^2 \qquad (3.42)$$

where $\Delta T = (T - T_a)$.

At this point, one might be tempted to simply estimate all the uncertainties as 1%, and Eq. (3.42) would then yield the uncertainty in the result, h, as 2%. This example, however, provides us with an opportunity to discuss several important practical points about using uncertainty analysis in planning experiments.

When an expression for the fractional uncertainty in a temperature (U_T/T) is encountered (as in Examples 3.1, 3.2, and 3.3), the temperature T *must* be expressed in absolute units of degrees Kelvin or Rankine and not in degrees Celsius or Fahrenheit. Likewise, the pressure p in (U_p/p) must be absolute, not gauge. However, an uncertainty in a quantity has the dimensions of a difference of two values of that quantity, and for U_T this means the units are Celsius degrees (which are equal to Kelvin degrees) or Fahrenheit degrees (which are equal to Rankine degrees). Thus, if someone says he can measure a temperature of 27°C with 1% uncertainty, U_T is 3°C (which is 0.01 times 300 K) and not 0.27°C. In general, percentage specifications for uncertainties in temperature measurements are easily misinterpreted and should be avoided.

In Eq. (3.42), we are working with a temperature difference and have the quantity $(U_{\Delta T}/\Delta T)$. If we estimate this as 1%, then we are saying the uncertainty is 0.2°C when ΔT is 20°C, the nominal value planned for the steady-state tests. However, if for some reason a test were run with $\Delta T = 5°C$, then our 1% specification indicates the uncertainty is 0.05°C, which is certainly overly optimistic for most experimental programs. The point to be emphasized here is that it is easier to estimate uncertainties on a fractional (or percentage) basis, but if a range of values of a variable is of interest or is likely to be encountered, estimating the uncertainty in that variable on an absolute basis gives less chance for misinterpretation of the behavior of the experiment.

The uncertainty in the power measurement in the steady-state technique is just such a case. Since we are interested in determining h for a range of air velocities, we would anticipate encountering a range of values for h. From Eq.

(3.38), we would then expect a range of values for the power W. If we estimate the uncertainty in the measurement of W as 2%, say, then U_W is 2 W when $W = 100$ W and 0.02 W when $W = 1$ W. The authors have found that it is, in general, productive to investigate a range of assumed absolute uncertainties in the planning phase of an experiment—that this gives the best opportunity to observe the behavior of the uncertainty in the result. In this case, for instance, one might want to check the behavior of the uncertainty in h for several assumed values of the uncertainty in W.

From Eq. (3.42) we can see that if we estimate a value for U_W, then we must have a numerical estimate of W also before U_h/h can be calculated. The amount of power W necessary to maintain a ΔT of 20°C is dependent on the value of h [see Eq. (3.38)], so we must estimate the range of values we expect to encounter in the result h before we can proceed with the uncertainty analysis. Considering the range of air velocities of interest and published results for convective heat transfer from infinitely long cylinders, it is reasonable to consider a range of hs from 10 to 1000 W/m²-C (about 1.8 to 180 Btu/hr-ft²-F). Putting numerical values into Eq. (3.38), this corresponds to values of input power W from 2.4 to 243 W.

We now need to estimate the uncertainties in D, L, ΔT, and W. Using

$$U_D = 0.025 \text{ mm } (0.001 \text{ in.})$$

$$U_L = 0.25 \text{ mm } (0.010 \text{ in.})$$

should be of the right order for the measurements of the cylinder dimensions. For the differential temperature measurement, let us initially consider

$$U_{\Delta T} = 0.25°C \text{ and } 0.5°C$$

For the power measurement, which will range up to 243 W at the highest h assumed, estimates of

$$U_W = 0.5 \text{ W and 1 W}$$

correspond to 0.2 and 0.4% of full scale on a meter with a 250 W range. Substitution into Eq. (3.42) using the lower estimates gives

$$
\begin{array}{cccc}
W: & D: & L: & \Delta T:
\end{array}
$$

$$\left(\frac{U_h}{h}\right)^2 = \left(\frac{0.5 \text{ W}}{W}\right)^2 + \left(\frac{0.025 \text{ mm}}{25.4 \text{ mm}}\right)^2 + \left(\frac{0.25 \text{ mm}}{152 \text{ mm}}\right)^2 + \left(\frac{0.25°C}{20°C}\right)^2$$

$$= \left(\frac{0.5 \text{ W}}{W}\right)^2 + (0.01 \times 10^{-4}) + (0.03 \times 10^{-4}) + (1.6 \times 10^{-4})$$

$$= \left(\frac{0.5 \text{ W}}{W}\right)^2 + (1.64 \times 10^{-4}) \tag{3.43}$$

TABLE 3.1   Results of General Uncertainty Analysis for Steady-State Technique

h (W/m²-C)	W (W)	U_h/h (%)			
		$U_W = 0.5$ W		1.0 W	
		$U_{\Delta T} = 0.25°C$	0.50°C	0.25°C	0.50°C
10	2.4	20.8	20.8	41.7	41.7
20	4.9	10.3	10.5	20.5	20.6
50	12.	4.4	4.9	8.4	8.7
100	24.	2.4	3.3	4.4	4.9
200	49.	1.6	2.7	2.4	3.2
500	121.	1.3	2.6	1.5	2.7
1000	243.	1.3	2.5	1.3	2.6

It is immediately apparent that the effect of the uncertainties in the cylinder dimensions is negligible relative to the influence of $U_{\Delta T}/\Delta T$ and that the relative influence of the uncertainty in the power measurement will be greater at low power and less at high power. To obtain numerical values from Eq. (3.43), we use Eq. (3.37)

$$W = h(\pi DL)(T - T_a) \tag{3.37}$$

which gives, upon use of the nominal values of D, L, and ΔT,

$$W = (0.01213 \text{ m}^2)(20°C)h = 0.2426\,h \tag{3.44}$$

where W is in watts and h in W/m²-C.

We can now choose values of h throughout the range of interest, use Eq. (3.44) to obtain the corresponding values of the power W, and then use Eq. (3.43) to determine the uncertainty in the experimental result h. This has been done for the measurement uncertainties previously assumed, and the results of the analysis are shown in Table 3.1 and are plotted in Figure 3.12.

The results in the figure show that the assumed values for the uncertainties in the differential temperature measurement will satisfy the requirements and could even be relaxed somewhat. The uncertainties of 0.5 and 1 W in the power measurement are adequate at higher values of h, but do not satisfy the requirements at low and moderate h values. The uncertainty in the power measurement is the dominant uncertainty (for the assumptions made in this analysis) in the low and moderate h range, and would have to be reduced below 0.5 W for the requirement of about 5% uncertainty in h to be met.

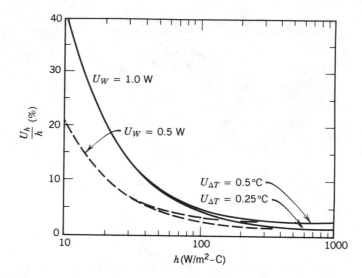

Figure 3.12   Uncertainty analysis results for the steady-state technique.

3-6.4   General Uncertainty Analysis: Transient Technique

The data reduction expression for the transient technique is given by Eq. (3.40)

$$h = \frac{Mc}{\pi D L \tau} \tag{3.40}$$

This is in the special product form of Eq. (3.20), so we can write down by inspection

$$\left(\frac{U_h}{h}\right)^2 = \left(\frac{U_\tau}{\tau}\right)^2 + \left(\frac{U_D}{D}\right)^2 + \left(\frac{U_L}{L}\right)^2 + \left(\frac{U_M}{M}\right)^2 + \left(\frac{U_c}{c}\right)^2 \tag{3.45}$$

The uncertainties in D and L have been estimated in the analysis for the steady-state case, and the same values will be used in this analysis. We should be able to measure the mass of the cylinder to within, say, 0.1 g, so we estimate

$$U_M = 0.0001 \text{ kg}$$

Rather than return to the literature to find some of the original data on the specific heat of aluminum, let us estimate the uncertainty in c as 1%. If the conclusions based on the results of our initial analysis turn out to be a strong function of this estimate, we can spend the time then to obtain a more defensible estimate. For the uncertainty in the measurement of the time corresponding to the time constant, it makes sense to see the effect of a range of values. For the initial analysis, we choose

$$U_\tau = 5, 10, \text{ and } 20 \text{ msec}$$

Substituting into Eq. (3.45) we find

$$\begin{array}{ccccc}
\tau: & D: & L: & M: & c: \\
\end{array}$$

$$
\left(\frac{U_h}{h}\right)^2 = \left(\frac{U_\tau}{\tau}\right)^2 + \left(\frac{0.025 \text{ mm}}{25.4 \text{ mm}}\right)^2 + \left(\frac{0.25 \text{ mm}}{152 \text{ mm}}\right)^2 + \left(\frac{0.0001 \text{ kg}}{0.020 \text{ kg}}\right)^2 + (0.01)^2
$$

$$
= \left(\frac{U_\tau}{\tau}\right)^2 + (0.01 \times 10^{-4}) + (0.03 \times 10^{-4})
$$

$$
+ (0.25 \times 10^{-4}) + (1 \times 10^{-4})
$$

$$
= \left(\frac{U_\tau}{\tau}\right)^2 + (1.29 \times 10^{-4}) \tag{3.46}
$$

The value of τ corresponding to an assumed value of h can be found by substituting the values of M, c, D, and L into Eq. (3.39)

$$\tau = \frac{Mc}{h(\pi DL)} \tag{3.39}$$

to obtain

$$\tau = \frac{148.9}{h} \tag{3.47}$$

where τ is in seconds and h is in W/m^2-C.

In a manner similar to that used in the steady-state analysis, we can now choose a value of h, determine the corresponding value of τ from Eq. (3.47), and substitute into Eq. (3.46) to obtain the uncertainty in the result h. This has been done for the range of values assumed for U_τ, and the results are shown in Table 3.2 and are plotted in Figure 3.13.

TABLE 3.2   Results of General Uncertainty Analysis for Transient Technique

h	τ	U_h/h (%)		
(W/m²-C)	(sec)	$U_\tau = 0.005$ sec	0.010 sec	0.020 sec
10	14.9	1.1	1.1	1.3
20	7.4	1.1	1.1	1.3
50	3.0	1.2	1.2	1.4
100	1.5	1.2	1.3	1.8
200	0.74	1.3	1.8	3.0
500	0.30	2.0	3.5	6.8
1000	0.15	3.5	6.8	13.4

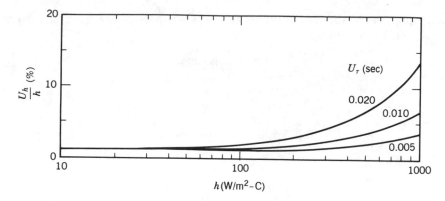

Figure 3.13   Uncertainty analysis results for the transient technique.

The results plotted in the figure show that the uncertainty in the experimental result increases with increasing h, which is opposite to the behavior seen for the steady-state technique. For the conditions and estimates used in this analysis, we can conclude that the transient technique will meet the requirements if the time constant can be measured with an uncertainty of 7 or 8 msec or less.

3-6.5   Implications of the Uncertainty Analysis Results

The application of general uncertainty analysis to the two proposed techniques for determining h has given us information on the behavior of the uncertainty in h that is not immediately obvious from the data reduction equations [Eqs. (3.38) and (3.40)]. For the steady-state technique, the percentage uncertainty in h increases as h decreases (that is, at lower air velocities). For the nominal values assumed, the uncertainty in the power measurement dominates for lower hs, whereas the uncertainty in the differential temperature measurement dominates for higher hs.

For the transient technique, the percentage uncertainty in h increases as h increases (i.e., at higher air velocities). For the nominal values assumed, the uncertainty in the measurement of the time constant dominates for higher hs, whereas the uncertainty in the value of the specific heat of the cylinder dominates for lower hs.

Since the trends of the uncertainty in h differ for the two techniques, there is a range of h for which the steady-state technique gives "better" results and a range for which the transient technique gives "better" results. The crossover point at which the techniques give equal results from an uncertainty standpoint depends on the estimates of the measurement uncertainties. Shown in Figure 3.14 is a comparison of the uncertainty analysis results for the two

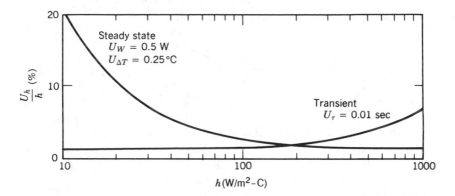

Figure 3.14 Comparison of uncertainty analysis results for the two proposed techniques.

techniques with

$$U_W = 0.5 \text{ W}$$

$$U_{\Delta T} = 0.25°\text{C}$$

$$U_\tau = 0.010 \text{ sec}$$

and with the other uncertainties as estimated previously. For this case, the transient technique is preferable for cases in which $h < 150$, the steady-state technique is preferable for cases in which $h > 200$, and the two techniques give results with about the same uncertainty for $150 < h < 200$ (where the units of h are W/m²-C).

One qualification on these conclusions must be noted. In free (or natural) convection, h is a function of $(T - T_a)$ and is not constant during the cooling process of the transient technique. The assumptions used in obtaining the data reduction expression [Eq. (3.40)] for the transient technique are therefore violated in cases in which the heat transfer is by free convection. These cases occur when the forced air velocity approaches zero.

3-7 NUMERICAL APPROXIMATION TO GENERAL UNCERTAINTY ANALYSIS

Recalling from Section 3-2, if an experimental result is given by a data reduction equation

$$r = r(X_1, X_2, \ldots, X_J) \tag{3.2}$$

then the uncertainty in the result can be expressed as

$$U_r = \left[\left(\frac{\partial r}{\partial X_1} U_{X_1} \right)^2 + \left(\frac{\partial r}{\partial X_2} U_{X_2} \right)^2 + \cdots + \left(\frac{\partial r}{\partial X_J} U_{X_J} \right)^2 \right]^{1/2} \quad (3.3)$$

There are cases when the data reduction expression is very complex and the task of obtaining the partial derivatives in Eq. (3.3) is extremely laborious. When a case is encountered such that the chances of obtaining correct analytical expressions for the partial derivatives is remote or, indeed, the expression (3.3) is so intimidating that one might be tempted to proceed without performing an uncertainty analysis, numerical approximations to the partial derivatives can be used. Using a forward differencing finite-difference approach, we can write

$$\frac{\partial r}{\partial X_1} \bigg|_{X_2, \ldots, \, X_J \text{ const}} = \frac{\Delta r}{\Delta X_1}$$

$$= \frac{r_{X_1 + \Delta X_1, \, X_2, \ldots, \, X_J} - r_{X_1, \, X_2, \ldots, \, X_J}}{\Delta X_1} \quad (3.48)$$

with similar expressions for the derivatives with respect to X_2 and X_J. The finite-difference approximation to Eq. (3.3) then becomes

$$U_r^2 \approx \left(\frac{\Delta r}{\Delta X_1} U_{X_1} \right)^2 + \left(\frac{\Delta r}{\Delta X_2} U_{X_2} \right)^2 + \cdots + \left(\frac{\Delta r}{\Delta X_J} U_{X_J} \right)^2 \quad (3.49)$$

This type of approach is fairly easily implemented using a spreadsheet program.

3-8 SUMMARY

In this chapter, we have seen how to use general uncertainty analysis in the planning phase of an experiment.

In the planning phase of an experimental program, we are trying to evaluate the approaches that might be used to find the answer(s) to the question that has been posed. What we usually do at this stage is to identify possible data reduction expressions and then evaluate them using general uncertainty analysis. The results of the uncertainty analysis are studied to see if the experimental result can be determined within an acceptable uncertainty and, if so, what the most critical measurements are from an uncertainty standpoint.

Sometimes, as seen in Section 3-5.4, the uncertainty analysis shows that a potential approach cannot meet the requirements (in terms of the maximum allowable uncertainty in the result) over the range of conditions expected. In such cases, the results of the uncertainty analysis can be used to investigate changes in the experimental approach that might lead to satisfactory results (as seen in Section 3-5.5). If no such changes are found, then either another experimental approach should be found or the effort abandoned. In either instance, the uncertainty analysis saves both time and money compared with the case in which no uncertainty analysis is done in the planning phase and the experiment is built and performed before the results are found to be unsatisfactory.

The point to be emphasized is that *an uncertainty analysis should always be performed in the planning phase of an experiment*. The amount of time and money spent on such an analysis is very small compared to the benefits obtained and insights gained, and pales beside the amount wasted on inappropriate experiments that have generated reams of useless results over the years.

REFERENCES

1. *Measurement Uncertainty*, ANSI/ASME PTC 19.1-1985 Part 1, 1986.
2. Shortley, G., and Williams, D., *Elements of Physics*, 4th ed., Prentice-Hall, Englewood Cliffs, NJ, 1965.
3. Hodkinson, J. R., "The Optical Measurement of Aerosols," in *Aerosol Science* (ed. C. N. Davies), Academic Press, New York, 1966, pp 287–357.
4. Ariessohn, P. C., Eustis, R. H., and Self, S. A., "Measurements of the Size and Concentration of Ash Droplets in Coal-Fired MHD Plasmas," *Proc. 7th Int. Conf. MHD Elec. Power Generation*, Vol II, 1980, pp 807–814.
5. Holve, D., and Self, S. A., "Optical Measurements of Mean Particle Size in Coal-Fired MHD Flows," *Combustion & Flame*, Vol 37, 1980, pp 211–214.
6. Moffat, R. J., "Using Uncertainty Analysis in the Planning of an Experiment," *J. Fluids Engineering*, Vol 107, 1985, pp 173–178.
7. Incropera, F. P., and Dewitt, D. P., *Fundamentals of Heat Transfer*, John Wiley, New York, 1981.

CHAPTER 3

Problems

3.1 Show by direct application of Eq. (3.15) to the data reduction equation

$$r = k(X_1)^a (X_2)^b (X_3)^c$$

that Eq. (3.21) is obtained. Consider a, b, c, and k to be constants that may be positive or negative.

3.2   The ideal gas equation of state can be written as

$$pV = mRT$$

For the nominal conditions and 95% confidence uncertainty estimates given below, what is the uncertainty in the volume of the gas?

$$p = 820 \pm 15 \text{ kPa}$$

$$m = 2.00 \pm 0.02 \text{ kg}$$

$$T = 305 \pm 3 \text{ K}$$

$$R = 0.287 \text{ kJ/kg K (assume known perfectly)}$$

3.3   The velocity V of a body traveling in a circular path can be measured with an uncertainty estimated at 4%. The radius R of the circular path can be measured within 2%. What is the uncertainty associated with determination of the normal acceleration, a_n, from

$$a_n = RV^2$$

Would it make much difference if the radius could be measured within 1%? 0.5%?

3.4   A column of length L and square cross-section ($b \times b$) is clamped at both ends. It has been proposed that the elastic modulus E of the column material be determined by axially loading the column until it buckles and using

$$p_{\text{cr}} = \frac{4\pi^2 EI}{L^2}$$

where p_{cr} is the buckling load, and the moment of inertia, I, is given by

$$I = \frac{1}{12}b^4$$

How well can E be determined if b, L, and p_{cr} are measured to within 1%? What is the relative importance of the uncertainties in the measured variables?

3.5   A refrigeration unit extracts energy from a cold space at T_c and transfers energy as heat to a warm environment at T_h. A Carnot refrigerator defines the ideal case, and its coefficient of performance is defined by

$$\text{C.O.P.} = \frac{T_c}{T_h - T_c}.$$

Assume equal uncertainties (in °C) in the measurement of T_c and T_h. How well must the temperature measurements be made for this ideal C.O.P. to be determined to within 1% when the nominal values of T_c and T_h are $-10°C$ and $20°C$?

3.6 A venturi flowmeter is to be used to measure the flow of air at low velocities, with air mass flow rate being given by

$$\dot{m} = CA\left(\frac{2p_1}{RT_1}(p_1 - p_2)\right)^{1/2}$$

where C is an empirically determined discharge coefficient, A is the throat flow area, p_1 and p_2 are the upstream and downstream pressures, T_1 is the absolute upstream temperature and R is the gas constant for air. We want to calibrate this meter and to determine C to within 2%. For the nominal values and estimated uncertainties below

$$p_1 = 30 \pm 0.5 \text{ psia}$$
$$T_1 = 70 \pm 3°F$$
$$\Delta p = p_1 - p_2 = 1.1 \pm 0.007 \text{ psia (measured directly)}$$
$$A = 0.75 \pm 0.001 \text{ in}^2$$

what is the allowable percentage uncertainty in the $\dot{m}$ measurement? That is, "how good" must the mass flow meter used in the calibration process be?

3.7 For the same nominal conditions as in the previous problem, the venturi is used to determine the air mass flow rate. If the value of C is known $\pm 1\%$, what will be the uncertainty in the determination of $\dot{m}$?

3.8 A heat transfer experiment is to be run with the anticipated behavior given by

$$\overline{T} = e^{-t/\tau}$$

where $\overline{T}$ is a nondimensional temperature that varies from 0 to 1 and can be measured $\pm 2\%$, t is time in seconds and can be measured $\pm 1\%$, and τ is the time constant that is to be determined from the experiment. Determine the behavior of the uncertainty in τ over the range of possible values for $\overline{T}$. Plot U_τ/τ vs. $\overline{T}$ for $\overline{T}$ from 0 to 1.

3.9 For the radial conduction of heat through a pipe wall, the thermal resistance (per unit length) is given by

$$R_{th} = \frac{\ln(r_2/r_1)}{2\pi k}$$

where r_1 and r_2 are the radii of the inner and outer surfaces, respectively, of the pipe wall and k is the thermal conductivity of the pipe material. Perform a general uncertainty analysis and then a parametric study to find the sensitivity of the uncertainty in R_{th} to the radius ratio r_2/r_1. (You might want to initially assume 1% uncertainties for the values of r_1, r_2, and k).

4

DESIGNING AN EXPERIMENT: DETAILED UNCERTAINTY ANALYSIS

In this chapter, we turn our attention to detailed uncertainty analysis. Using this approach, which is consistent with that outlined in the Standard [1], the details of the bias and the precision errors in each measured variable are considered, and the propagation of the bias and precision limits into the experimental result are investigated separately. This is in contrast to the general uncertainty analysis used in the planning phase of an experiment, which was discussed in the previous chapter.

The primary reason for considering the more complex approach of detailed uncertainty analysis is that it is very useful in the design, construction, debugging, data analysis, and reporting phases of an experiment to consider separately the bias and precision components of uncertainty. The bias is a fixed error that can be reduced by calibration. However, the precision error is a variable error that can be reduced by the use of multiple readings. This differing behavior of the two components of the uncertainty makes it desirable and necessary to consider the components separately. In the following presentation, we will be using the bias and precision limits to represent these errors. Referring back to the introduction of these concepts in Chapter 2, we are reminded that these limits represent the 95% confidence interval estimates for the bias error and the precision error in a measurement.

Although we used the word "complex" in the previous paragraph in connection with detailed uncertainty analysis, this should not be cause for concern or a hesitancy to use the approach. As shown in the following sections of this chapter, the application of detailed uncertainty analysis is accomplished through a series of logical steps that are themselves fairly straightforward.

4-1   DETAILED UNCERTAINTY ANALYSIS: OVERVIEW

The situation that we wish to analyze is illustrated in Figure 4.1, which shows a flow diagram of the propagation of errors into an experimental result. Each of the measurement systems that is used to measure the value of an individual variable X_i is influenced by a large number of elemental error sources. The effects of these elemental errors are manifested as a bias error limit B_i and a precision error limit P_i in the measured value of the variable. These errors in the measured values then propagate through the data reduction equation and yield the bias and precision errors in the experimental result.

The procedure in detailed uncertainty analysis is to investigate the contributions of the elemental error sources, obtain estimates of the bias and precision limits for each measured variable, and use the uncertainty analysis expression to obtain values for the bias limit B_r and precision limit P_r of the experimental result. The uncertainty in the result is then expressed by combining these two error components in one of two ways:

$$U_{r_{RSS}} = \left[B_r^2 + P_r^2 \right]^{1/2} \tag{4.1}$$

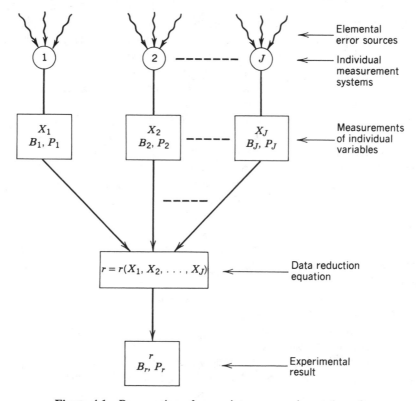

Figure 4.1   Propagation of errors into an experimental result.

or

$$U_{r_{\text{ADD}}} = B_r + P_r \qquad (4.2)$$

The root-sum-square uncertainty, $U_{r_{\text{RSS}}}$, results in approximately 95% coverage of the true value. The additive uncertainty, $U_{r_{\text{ADD}}}$, results in approximately 99% coverage when the bias and precision contributions are of the same order and 95% coverage when one is negligible relative to the other [1]. These statements assume that the bias limit estimates have all been made at 95% confidence (20 to 1 odds) and that t in the precision limit ($P = tS$) has been chosen for a 95% confidence level. We will always use RSS combination and 95% confidence levels in this book unless specifically stated otherwise.

In the sections that follow in this chapter, we will discuss methods for determining the bias and precision limits in measurements of individual variables and the way in which they propagate into experimental results.

4-2 PROPAGATION OF BIAS ERRORS INTO AN EXPERIMENTAL RESULT

In all of our discussions of bias, we will assume that corrections have been made for all of the bias errors whose values are known, such as those determined by calibration. This means that all remaining bias errors must be estimated. Because the "true" value of a measured variable is never known and because bias is a fixed error, there are no measurement or statistical procedures such as those of Chapter 2 that can be used to provide values of the biases in measurements. (It might be recalled here that biases can be determined by calibration, and thus eliminated, only to a certain degree. The bias associated with the reference or standard used in the calibration procedure will remain in the output of the instrument that is calibrated.)

Shown schematically in Figure 4.2 are the steps in the procedure for determining B_r, the bias limit in the experimental result. Each measurement system used to determine the value of an individual variable is influenced by biases from a number of elemental error sources. These biases are estimated and combined to form the estimate of the bias limit for each measured variable. The bias limits of the individual variables are then propagated through the uncertainty analysis expression to obtain the bias limit for the experimental result.

4-2.1 Biases from Elemental Error Sources

The first step is to consider the biases due to the elemental error sources that affect each of the measurements. The Standard [1] suggests that these sources

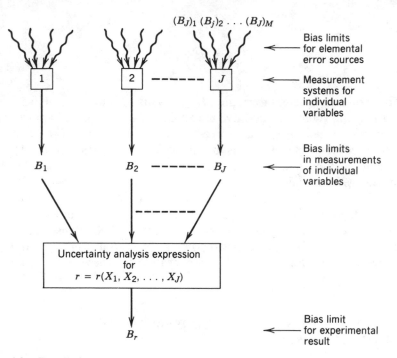

Figure 4.2   Detailed uncertainty analysis—determination of bias limit for experimental result.

be grouped into the categories of

a. calibration
b. data acquisition
c. data reduction

Some bias always remains in an instrument calibration as no reference or standard is perfect. Biases in the data acquisition category include environmental and installation effects on the transducer as well as the biases in the system that acquires, conditions, and stores the output of the transducer. Biases in the data reduction category include those in curvefits and in computational resolution.

In considering the measurement system for the Jth variable (as in Figure 4.2), we might identify and estimate 95% confidence values of a total of M biases in the three categories listed and then have

$$(B_J)_1, (B_J)_2, \ldots, (B_J)_M$$

as the bias limits from elemental sources that contribute to the bias in the value of the Jth measured variable, X_J.

4-2.2   Combining Elemental Bias Limits

Once all of the elemental bias limits that are thought to be significant have been estimated, they are combined using RSS to obtain the bias limit for each measured variable. For the Jth variable influenced by biases from M significant elemental error sources,

$$B_J = \left[(B_J)_1^2 + (B_J)_2^2 + \cdots + (B_J)_M^2\right]^{1/2}$$

$$= \left[\sum_{k=1}^{M} (B_J)_k^2\right]^{1/2} \tag{4.3}$$

On first thought, it might seem that the bias limits should be added together rather than being combined by RSS. However, direct addition would be a good estimate only for the case in which all elemental bias limits were of the same sign and also at their maximum values. Since this is improbable, the RSS combination is used [1] to obtain the 95% confidence estimate of the bias in the measurement.

4-2.3   Propagation of Bias Limits into the Experimental Result

Once the bias limits have been estimated for each of the measured variables in the data reduction equation

$$r = r(X_1, X_2, \ldots, X_J) \tag{4.4}$$

the bias limit for the experimental result is found from the uncertainty analysis expression

$$B_r = \left[\left(\frac{\partial r}{\partial X_1} B_1\right)^2 + \left(\frac{\partial r}{\partial X_2} B_2\right)^2 + \cdots + \left(\frac{\partial r}{\partial X_J} B_J\right)^2\right]^{1/2} \tag{4.5}$$

The development of Eq. (4.5) is presented and discussed in Appendix B. It is assumed that the relationship given by (4.4) is continuous and has continuous derivatives in the domain of interest and that the bias limits B_i for the measured variables are independent of one another. Situations in which this last assumption is not valid are considered in the next section (4-2.4).

As discussed in Chapter 3, if the partial derivatives are defined as absolute sensitivity coefficients so that

$$\theta_i = \frac{\partial r}{\partial X_i} \tag{4.6}$$

then Eq. (4.5) can be written as

$$B_r = \left[\sum_{i=1}^{J} \theta_i^2 B_i^2 \right]^{1/2} \tag{4.7}$$

Alternatively, Eq. (4.5) can be divided by the result r and recast into the form

$$\left(\frac{B_r}{r} \right)^2 = \left(\frac{1}{r} \frac{\partial r}{\partial X_1} B_1 \right)^2 + \left(\frac{1}{r} \frac{\partial r}{\partial X_2} B_2 \right)^2 + \cdots + \left(\frac{1}{r} \frac{\partial r}{\partial X_J} B_J \right)^2 \tag{4.8}$$

which will in most cases yield the algebraically simplest formulation.

The partial derivatives in the above expressions are the same ones that have been determined for use in the general uncertainty analysis performed in the planning phase of the experiment (see Chapter 3).

4-2.4   Procedures for Nonindependent Bias Limits

In the previous section, the equation we used for propagation of the bias limits in individual measurements into the bias limits of the experimental result [Eq. (4.5)] was based on the assumption that the bias limit in the measurement of each individual variable was independent of the bias limit in each of the other measured variables. There are important practical cases, however, in which the bias limits in the measurements of different individual variables are not independent of one another. Typical instances are those in which two variables are measured with the same transducer (the fluid temperature at different points in a flow field measured with the same probe at different times) and in which two transducers have been calibrated against the same standard.

In these cases in which some of the bias limits of the measurements are correlated (or not independent), the uncertainty analysis expression for propagation of the bias limits of individual measurements into the experimental result is (from Appendix B)

$$B_r^2 = \sum_{i=1}^{J} \left[\theta_i^2 B_i^2 + \sum_{k=1}^{J} \theta_i \theta_k \rho_{ik} B_i B_k (1 - \delta_{ik}) \right] \tag{4.9}$$

where the Kronecker delta is

$$\delta_{ik} \begin{cases} = 1 & i = k \\ = 0 & i \neq k \end{cases} \tag{4.10}$$

the coefficient of correlation between the biases in X_i and X_k is

$$\rho_{ik} = \rho_{ki} \tag{4.11}$$

and, as previously,

$$\theta_i = \frac{\partial r}{\partial X_i} \tag{4.6}$$

Equation (4.9) can appear to be somewhat intimidating—particularly to the person inexperienced in uncertainty analysis who is already worried about how to estimate the B_is in the first place. In actual practice, however, the specific form of the general equation (4.9) that is used is considerably simpler.

As an example, consider a case in which the experimental result r is related to three variables (x, y, and z) through the data reduction equation

$$r = r(x, y, z) \tag{4.12}$$

The uncertainty analysis equation for the bias limit of the result is, using Eq. (4.9),

$$B_r^2 = \left(\frac{\partial r}{\partial x}\right)^2 B_x^2 + \left(\frac{\partial r}{\partial y}\right)^2 B_y^2 + \left(\frac{\partial r}{\partial z}\right)^2 B_z^2$$

$$+ 2\left(\frac{\partial r}{\partial x}\right)\left(\frac{\partial r}{\partial y}\right)\rho_{xy}B_x B_y + 2\left(\frac{\partial r}{\partial x}\right)\left(\frac{\partial r}{\partial z}\right)\rho_{xz}B_x B_z$$

$$+ 2\left(\frac{\partial r}{\partial y}\right)\left(\frac{\partial r}{\partial z}\right)\rho_{yz}B_y B_z \tag{4.13}$$

On the right-hand side of this equation, the first three terms are those with which we are already familiar. The final three terms in the equation take into account effects that occur if the biases in the measurements of x, y, and z are not independent of one another, that is, if the biases in the measurements of x, y, and/or z are *correlated* with one another.

For the sake of this discussion, assume that only the biases in the x and y measurements are correlated, so

$$\rho_{xz} = \rho_{yz} = 0$$

and the final two terms in (4.13) are zero. In practice, the remaining correlation term can be rewritten as

$$2\left(\frac{\partial r}{\partial x}\right)\left(\frac{\partial r}{\partial y}\right)\rho_{xy}B_x B_y = 2\left(\frac{\partial r}{\partial x}\right)\left(\frac{\partial r}{\partial y}\right)(1) B_x' B_y' \tag{4.14}$$

where the correlation coefficient ρ_{xy} has been set equal to one and B_x' and B_y' are the portions of B_x and B_y that arise from identical error sources (and are presumably perfectly correlated). The normalized uncertainty analysis equation for the bias limit for this case in which the biases in only two measurements (X_1 and X_2) are correlated would then have the form

$$\left(\frac{B_r}{r}\right)^2 = \left(\frac{1}{r}\frac{\partial r}{\partial X_1}B_{X_1}\right)^2 + \left(\frac{1}{r}\frac{\partial r}{\partial X_2}B_{X_2}\right)^2 + \cdots + \left(\frac{1}{r}\frac{\partial r}{\partial X_J}B_{X_J}\right)^2$$
$$+ \frac{2}{r^2}\frac{\partial r}{\partial X_1}\frac{\partial r}{\partial X_2}B'_{X_1}B'_{X_2} \tag{4.15}$$

The following example illustrates the application of the ideas previously discussed.

Example 4.1

Two plastic-encased thermistor probes are to be used to measure the inlet and outlet temperatures of water that flows through a heat exchanger (Figure 4.3). We wish to determine the bias in the temperature difference

$$\Delta T = T_2 - T_1$$

The probes are guaranteed by the manufacturer to match the "standard" resistance–temperature curve for this particular type of thermistor within 1.0°C over the temperature range of interest to us. We assume that this is a bias specification, so that averaging multiple readings would not reduce this number.

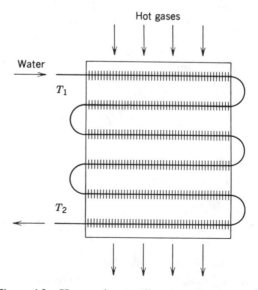

Figure 4.3 Heat exchanger diagram—Example 4.1.

If this amount of bias is unsatisfactory for our purposes, we can calibrate the two probes by immersing them in a benchtop constant temperature water or oil bath and using a calibrated thermometer (with 0.3°C bias) as a laboratory standard. The constant temperature bath has maximum spatial temperature nonuniformity of ± 0.1°C, according to the manufacturer, and we assume this value is a reasonable 95% confidence bias estimate.

After discussion with the heat exchanger manufacturer, we estimate (at 95% confidence) the biases due to water temperature nonuniformities at the inlet and outlet pipe cross sections where the probes will be placed as 0.1°C at the inlet and 0.2°C at the outlet. These biases account for the fact that the probes will probably not "see" the average water temperature at a cross section since the water temperature is not uniform across the cross section.

Biases from all other elemental error sources are considered to be negligible compared to those enumerated in the previous paragraphs.

A tabulation of the possible elemental bias sources for the two variables that will be measured in this experiment gives

	Bias	
	---	---
Elemental Source	T_1 (°C)	T_2 (°C)
1. As-delivered probe specification	1.0	1.0
2. Reference thermometer—calibration	0.3	0.3
3. Bath nonuniformity—calibration	0.1	0.1
4. Spatial variation—data acquisition	0.1	0.2

We now can use Eq. (4.3) and combine the bias limits from the elemental sources 1 and 4 to obtain

$$B_{T_1} = \left[(1.0)^2 + (0.1)^2\right]^{1/2} = 1.0°C$$

$$B_{T_2} = \left[(1.0)^2 + (0.2)^2\right]^{1/2} = 1.0°C$$

as the bias limits for the two measured variables when we do not calibrate the probes. Similarly, we combine the bias limits from the elemental sources 2, 3, and 4 to obtain

$$B_{T_1} = \left[(0.3)^2 + (0.1)^2 + (0.1)^2\right]^{1/2} = 0.33°C$$

$$B_{T_2} = \left[(0.3)^2 + (0.1)^2 + (0.2)^2\right]^{1/2} = 0.37°C$$

as the bias limits for the two measured variables when we do calibrate the probes.

The data reduction equation in this case is

$$\Delta T = T_2 - T_1$$

and the uncertainty analysis expression [from Eq. (4.13) and incorporating Eq. (4.14)] is

$$B_{\Delta T}^2 = \theta_{T_1}^2 B_{T_1}^2 + \theta_{T_2}^2 B_{T_2}^2 + 2\theta_{T_1}\theta_{T_2} B_{T_1}' B_{T_2}'$$

where

$$\theta_{T_1} = \frac{\partial \Delta T}{\partial T_1} = -1$$

$$\theta_{T_2} = \frac{\partial \Delta T}{\partial T_2} = 1$$

For the case in which we do not calibrate the probes, the biases affecting the probes are independent of one another and

$$B_{T_1}' = B_{T_2}' = 0$$

so that

$$B_{\Delta T}^2 = (1)(1.0)^2 + (1)(1.0)^2$$
$$B_{\Delta T} = 1.4°C$$

Now for the case in which we calibrate the probes, the uncertainty analysis expression becomes

$$B_{\Delta T}^2 = (1)(0.33)^2 + (1)(0.37)^2 + (2)(-1)(1) B_{T_1}' B_{T_2}'$$

We will consider three possibilities:

a. If the probes were to be calibrated at different times against two different reference thermometers, then

$$B_{T_1}' = B_{T_2}' = 0$$

as there would be no correlation between any of the elemental biases for T_1 and T_2 and therefore

$$B_{\Delta T} = 0.50°C$$

b. If the probes were to be calibrated at the same time against the same thermometer but were placed at different locations in the "constant temperature" bath, then the bias of the thermometer would affect each of the probes the same but the bias of the bath nonuniformity would not. In this case,

$$B'_{T_1} = B'_{T_2} = 0.3°C$$

and we obtain

$$B^2_{\Delta T} = (1)(0.33)^2 + (1)(0.37)^2 + (2)(-1)(1)(0.3)(0.3)$$
$$= 0.1089 + 0.1369 - 0.1800$$

or

$$B_{\Delta T} = 0.26°C$$

c. If the probes were to be calibrated at the same time against the same thermometer and were placed adjacent to each other at essentially the same location in the "constant temperature" bath, then the bias of the thermometer and the bias of the bath nonuniformity, sources 2 and 3, would affect each of the probes the same and would thus be correlated. In this case,

$$B'_{T_1} = B'_{T_2} = \left[(0.3)^2 + (0.1)^2\right]^{1/2} = 0.32$$

so that

$$B^2_{\Delta T} = (1)(0.33)^2 + (1)(0.37)^2 + (2)(-1)(1)(0.32)(0.32)$$
$$= 0.1089 + 0.1369 - 0.2048$$

or

$$B_{\Delta T} = 0.20°C$$

This example illustrates the importance of considering the effect of correlated bias limits when determining the bias in an experimental result. In this particular instance, proper design of the calibration procedure allowed the bias in the temperature difference (0.20°C) to be less than the bias (0.33°C or 0.37°C) in either of the measured temperatures.

4-2.5 Determining Bias Limits

As previously discussed, the 95% confidence interval bias limits must always be estimated. No matter how large a data sample might be, there is no way to directly calculate the bias errors. Information on bias errors can be inferred from comparison of independent measurements that depend on different physical principles or that have been independently calibrated. Other bases for estimates are previous experience of the experimenter and of others, instrument manufacturer's information and specifications, and comparison of measurements with known values (such as limiting cases, etc.).

Bias Error Sources. As presented in Section 4-2.1, bias errors are generally grouped into the categories of calibration, data acquisition, and data reduction. The calibration process should always be done with the measurement system (transducer, signal conditioner, data recording device, etc.) as close to the actual measurement condition and test installation arrangement as possible. The bias associated with the calibrated measurement system can then be reduced until it approaches the bias in the calibration standard.

In some cases, because of time or cost, the measurement system is not calibrated in its test configuration. In these cases the bias errors that are inherent in the installation process must be included in the overall bias determination. Moffat [2] points out that these installation errors include interactions of the transducer with the system (such as radiation errors in temperature measurements) and system disturbances because of the presence of the transducer. These biases can sometimes be accounted for by modifying the data reduction equation to include these effects. The bias estimates for the new terms in the modified data reduction equation then replace the estimates that would be made for the installation bias errors.

Another point presented by Moffat is the idea of conceptual bias errors. Are we really measuring the variable that is required in the data reduction equation? For instance, in fluid flow in a pipe, we might need the average velocity but only be able to measure the velocity at one point with the equipment available. The relationship between this single measurement of the velocity and the average velocity must be inferred from auxiliary information and included as a contributor to the bias in the uncertainty calculation.

With all of these possible sources of bias error, one might decide that it is impossible to make accurate measurements. This is not the case. With experience in making specific measurements, one learns which errors are important and which errors are negligible. The key point is to think through the entire measurement process to properly account for the biases.

Uncertainty in Property Values. In many experiments, values for some of the variables in the data reduction equations are not measured, but rather are found from reference sources. This is often the case for material properties, which are typically tabulated as a function of temperature. Whether we enter the table to obtain a property value or use a curvefit equation that represents the table, we obtain the same particular property value each time we use a given temperature. This value is not the true value—it is a "best estimate" based on experimental data and has an uncertainty associated with it.

As a specific example, consider that we need values for some thermal properties of air, say specific heat and thermal conductivity. Shown in Figures 4.4 and 4.5 are experimental data on specific heat and thermal conductivity, respectively, over a range of temperature. These are presented in the form of a percentage departure of the experimental data from the values tabulated in the National Bureau of Standards publication [3]. From these figures, it should be obvious that assuming the bias limits and precision limits of property values

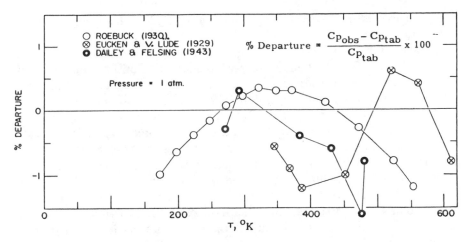

Figure 4.4 Departures of low-pressure experimental specific heats from the tabulated values for air (from U.S. National Bureau of Standards Circular 564, 1955).

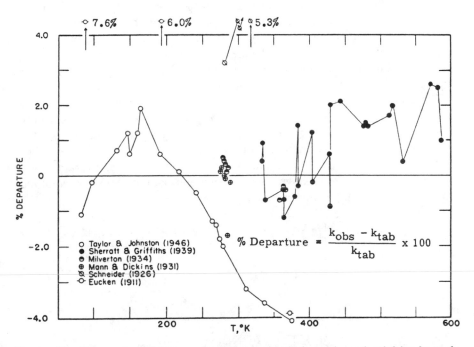

Figure 4.5 Departures of low-pressure experimental thermal conductivities from the tabulated values for air (from U.S. National Bureau of Standards Circular 564, 1955).

are negligible is not generally a good assumption, even for such a common substance as air.

How should the precision limit and bias limit be estimated for a property value obtained from a table or curvefit equation? Consider that once we choose a table or curvefit to use, we will always obtain *exactly* the same property value for a given temperature. If we entered the air specific heat table at 400 K 100 times over the period of a month, each of the 100 specific heat values would be the same—there would be a zero precision limit associated with the value that we read from the table. This is the case regardless of the amount of scatter in the experimental data on which the table is based.

We conclude then that all of the errors (both precision and bias) in the experimental property data are "fossilized" [2] into a bias error in values taken from the table or curvefit equation that represents the experimental data. We thus should take our best estimate of the overall uncertainty in a property value and use that as the fossilized bias limit for that property. In practical terms, this generally means estimating an uncertainty band based on the data scatter from different experiments, such as that seen in Figures 4.4 and 4.5.

Asymmetric Bias Errors. In most cases the bias errors will be those symmetrical fixed errors that remain around the true values of the measured variables after all calibrations and corrections have been taken into account. These biases are combined as shown in the previous sections to yield the bias limit of the result. The bias and precision limits are then combined by the RSS method [Eq. (4.1)] to yield the uncertainty of the result.

There are some situations, however, in which it will not be possible or practical to correct the measurements so that all the bias limits will be symmetrical. If asymmetrical bias errors are not or cannot be adjusted by modifying the data reduction equation, then the positive and negative components of the bias limits must be handled separately. For those variables with asymmetrical bias, the positive and negative components must be combined separately using Eq. (4.3) to calculate B_J^+ and then B_J^-.

$$B_J^+ = \left[(B_J^+)_1^2 + (B_J^+)_2^2 + \cdots + (B_J^+)_M^2 \right]^{1/2}$$

$$= \left[\sum_{k=1}^{M} (B_J^+)_k^2 \right]^{1/2} \tag{4.16}$$

and

$$B_J^- = \left[(B_J^-)_1^2 + (B_J^-)_2^2 + \cdots + (B_J^-)_M^2 \right]^{1/2}$$

$$= \left[\sum_{k=1}^{M} (B_J^-)_k^2 \right]^{1/2} \tag{4.17}$$

These components are then used with the bias errors in the other measured variables in Eq. (4.5) or Eq. (4.9) to obtain B_r^+ and B_r^- so that

$$(B_r^+)^2 = \sum_{i=1}^{J} \left[\theta_i^2 (B_i^+)^2 + \sum_{k=1}^{J} \theta_i \theta_k \rho_{ik} B_i^+ B_k^+ (1 - \delta_{ik}) \right] \qquad (4.18)$$

and

$$(B_r^-)^2 = \sum_{i=1}^{J} \left[\theta_i^2 (B_i^-)^2 + \sum_{k=1}^{J} \theta_i \theta_k \rho_{ik} B_i^- B_k^- (1 - \delta_{ik}) \right] \qquad (4.19)$$

where for the variables with symmetrical bias limits, $B_i^+ = B_i^-$. The procedure is then completed by using the RSS method [Eq. (4.1)] to combine the separate bias limits of the result with the precision limit to determine U_r^+ and U_r^-

$$U_r^+ = \left[(B_r^+)^2 + P_r^2 \right]^{1/2} \qquad (4.20)$$

and

$$U_r^- = \left[(B_r^-)^2 + P_r^2 \right]^{1/2} \qquad (4.21)$$

In the last section of this chapter and in the following chapters in which we discuss the design, construction, debugging, execution, data analysis, and reporting phases of experimental programs, we will consider detailed examples that will illustrate how estimates of bias limits might be logically obtained.

4-3 PROPAGATION OF PRECISION ERRORS INTO AN EXPERIMENTAL RESULT

Once the bias limit for the result has been determined using the approach discussed in Section 4-2, the precision limit (P_r) for the result must be determined before the uncertainty in the result can be found from either Eq. (4.1) or (4.2). In the general case, we will consider the experimental result that is determined from a single test using

$$r = r(X_1, X_2, \ldots, X_J) \qquad (4.4)$$

In such a test, a variable X_i might be measured only once or it might be measured repeatedly at a nominally constant set point so that an average value $\overline{X}_i$ is used in Eq. (4.4), but the result r is determined from (4.4) only once at a given experimental condition. Cases in which r is determined from multiple tests at the same experimental condition are considered in Section 4-3.3.

Recall that the precision limit for the measurement of a variable is the product of the precision index S and the factor t (determined from the t-distribution) such that the range from $-(tS)$ to $+(tS)$ about the measured value would contain the true value of the variable 95% of the time if all biases were zero. For the normally distributed error distributions that we are assuming, t can be taken as equal to two when S is determined based on a large ($N > 30$) sample size.

For the general case of an experimental result that is obtained from a single test, the precision limit of each variable X_i must be determined. These precision limits are determined by several methods such as repeated measurements, or auxiliary tests, or previous experience, or from estimates as with the readability of a scale. The determination of precision limits of individual measurements and their propagation into the experimental result are discussed in Sections 4-3.1 and 4-3.2. In Section 4-3.4, the special case is considered in which the precision index for each variable is statistically determined from repeated measurements with some of the measurements having small sample sizes ($N < 30$).

The general case procedure for determination of the precision limit for a result from a single test is illustrated schematically in Figure 4.6. Each measurement system used to determine the value of an individual variable is influenced by precision errors from a number of elemental error sources. These random errors combine to cause the precision errors in the measurement of each variable. These precision errors are quantified by determining the preci-

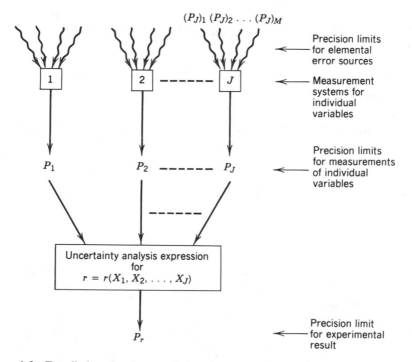

Figure 4.6 Detailed uncertainty analysis—determination of precision limit for experimental result. Result determined from a single test.

sion limit of each measured variable. The precision limits of the individual variables are then propagated through the uncertainty analysis expression to obtain the precision limit P_r for the experimental result.

4-3.1 Precision Limits of Individual Measurements

The way in which the precision limits of the measurements of the individual variables are determined will vary depending on both the particular experiment and the phase of the experiment with which we are concerned. For example, when we are determining which transducers and data acquisition equipment should be specified during the design phase of a completely new experiment, estimates must be made based on all the information available— our previous experience and that of others, manufacturers' specifications, etc. For cases in which a precision limit P_i is estimated, the estimate ($\pm tS$) should correspond to what we would expect to measure for large sample sizes ($N > 30$) so that t will equal two. Stated another way, the estimated precision limit should be that $+$ and $-$ band that contains the mean value of the variable with 95% confidence.

On the other hand, when in the execution phase of an experiment utilizing a test apparatus that has been used many times in the past, we may have previous measurements available with which to determine each of the P_is. In other cases, we may make repeated measurements of the individual variables and use these to calculate the P_is. However, in some cases a variable may have a precision error variation over a much longer time frame than that required to make the measurement. In these cases, the precision limit must be determined from auxiliary tests that cover the appropriate timeframe.

When there are several identifiable factors causing the precision error in a measured variable, it sometimes might be desirable and worthwhile to determine the precision limit by considering the contributions of the precision limits of the elemental error sources. This procedure is very similar to that discussed for the determination of a bias limit. If for the Jth variable, for instance, we are able to identify, isolate, and determine the precision limits from M significant elemental error sources, we have

$$(P_J)_1, (P_J)_2, \ldots, (P_J)_M$$

These are then combined using root-sum-square to obtain

$$P_J = \left[(P_J)_1^2 + (P_J)_2^2 + \cdots + (P_J)_M^2 \right]^{1/2}$$

$$= \left[\sum_{k=1}^{M} (P_J)_k^2 \right]^{1/2} \tag{4.22}$$

Once the precision limits for the measured variables have been either estimated or calculated, they are propagated into the result as described in the following.

4-3.2 Propagation of Precision Limits into the Experimental Result

Once the P_is have been determined for each of the measured variables in the data reduction equation

$$r = r(X_1, X_2, \ldots, X_J) \tag{4.4}$$

the precision limit for the experimental result is found from

$$P_r = \left[\left(\frac{\partial r}{\partial X_1} P_1 \right)^2 + \left(\frac{\partial r}{\partial X_2} P_2 \right)^2 + \cdots + \left(\frac{\partial r}{\partial X_J} P_J \right)^2 \right]^{1/2} \tag{4.23}$$

The development of Eq. (4.23) is presented and discussed in Appendix B. It is assumed that the relationship given by (4.4) is continuous and has continuous derivatives in the domain of interest and that the precision limits P_i for the measured variables are independent of one another. Since the precision limits are statistical measures of the effect of random errors and variations on a measurement, the assumption of independent precision limits in the individual measurements seems justified.

If absolute sensitivity coefficients θ_i as defined in Eq. (4.6) are used, then Eq. (4.23) can be written as

$$P_r = \left[\sum_{i=1}^{J} \theta_i^2 P_i^2 \right]^{1/2} \tag{4.24}$$

Alternatively, Eq. (4.23) can be divided by the result r and recast into the form

$$\left(\frac{P_r}{r} \right)^2 = \left(\frac{1}{r} \frac{\partial r}{\partial X_1} P_1 \right)^2 + \left(\frac{1}{r} \frac{\partial r}{\partial X_2} P_2 \right)^2 + \cdots + \left(\frac{1}{r} \frac{\partial r}{\partial X_J} P_J \right)^2 \tag{4.25}$$

which will in most cases yield the algebraically simplest formulation. It should again be noted that the partial derivatives in the previous expressions are the same ones that have been determined for use in the general uncertainty analysis done in the planning phase of the experiment.

Several comments must be made about the proper interpretation of the X_is in the data reduction equation and the corresponding P_is that are used in the uncertainty analysis equations. Recalling the discussion presented in Chapter 2, the appropriate precision limit to use with a variable X_i that is (or will be)

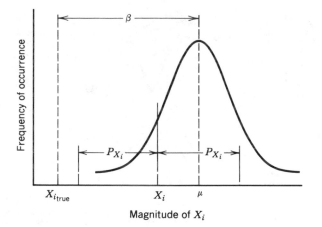

Figure 4.7 The 95% confidence precision limit interval around a single reading of a variable X_i.

determined with a single reading is the precision index of the sample population times the factor t

$$P_{X_i} = tS_{X_i} = t \left\{ \frac{1}{N-1} \sum_{k=1}^{N} [(X_i)_k - \overline{X}_i]^2 \right\}^{1/2} \qquad (4.26)$$

Of course for a single reading, P_{X_i} will have to be estimated or be available from previous measurements of the variable. This method of determining P_{X_i} also applies to the case in which precision errors with large time constants affect the reading.

As shown in Figure 4.7, the $\pm P_{X_i}$ band around the reading X_i contains the (biased) mean value of the measured variable with 95% confidence. Therefore, in the uncertainty analysis equations, X_i and P_i should be interpreted as

$$X_i \text{ and } P_{X_i}$$

when the value of X_i that is used in the data reduction equation (4.4) is determined from a single reading.

When the value of the variable is (or will be) determined as the mean ($\overline{X}_i$) of N separate readings, then the precision limit of the mean of the sample

$$P_{\overline{X}_i} = tS_{\overline{X}_i} = tS_{X_i}/\sqrt{N} \qquad (4.27)$$

should be used. As shown in Figure 4.8, the $\pm P_{\overline{X}_i}$ band around the sample mean $\overline{X}_i$ contains the (biased) mean value of the measured variable with 95% confidence. In the case in which the value of the variable that is used in the

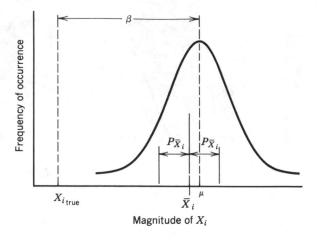

Figure 4.8 The 95% confidence precision limit interval around the mean value of a sample of readings of a variable X_i.

data reduction equation (4.4) is determined as the average of N separate readings, the values

$$\overline{X}_i \text{ and } P_{\overline{X}_i}$$

should be used in the uncertainty analysis equations.

Once the precision limit of the result is found from Eq. (4.23), the uncertainty in the result r is expressed as

$$U_{r_{\text{RSS}}} = \left[B_r^2 + P_r^2 \right]^{1/2} \tag{4.1}$$

where $r \pm U_{r_{\text{RSS}}}$ provides a 95% coverage interval of the true result.

4-3.3   Direct Determination of the Precision Limit of the Result: Multiple Tests

We now consider the type of experimental program in which multiple tests are performed using the same experimental apparatus and the final experimental result is determined as the mean of the results of the individual tests. In such a program, the process shown in Figure 4.1 for a single test is replicated M times, yielding individual values

$$r_1, r_2, \ldots, r_M$$

which are averaged to produce the "best" value of the result

$$\bar{r} = \frac{1}{M} \sum_{k=1}^{M} r_k \tag{4.28}$$

The bias limit associated with this average result is the same as the bias limit B_r for a single test and is found using the techniques discussed in Section 4-2. Because biases are fixed errors, they are not affected by averaging the results of multiple tests. The precision index of the sample population of the M individual test results is given by

$$\hat{S}_r = \left[\frac{1}{M-1} \sum_{k=1}^{M} (r_k - \bar{r})^2 \right]^{1/2} \tag{4.29}$$

and has $M - 1$ degrees of freedom.

The precision index of the average result, $\bar{r}$, from M tests is

$$S_{\bar{r}} = \frac{\hat{S}_r}{\sqrt{M}} \tag{4.30}$$

It also has $M - 1$ degrees of freedom, $\nu_{\bar{r}}$, and this value is used with the t-distribution to determine the t value corresponding to a 95% confidence level.

The uncertainty in the average result $\bar{r}$ is then expressed as

$$U_{\bar{r}_{RSS}} = \left[B_r^2 + (P_{\bar{r}})^2 \right]^{1/2} \tag{4.31}$$

for a 95% coverage or as

$$U_{\bar{r}_{ADD}} = B_r + P_{\bar{r}} \tag{4.32}$$

for a 99% coverage where

$$P_{\bar{r}} = t S_{\bar{r}} \tag{4.33}$$

We note here the obvious fact that the equations in this section cannot be applied until the experiment has been run and the results are in hand. Prior to the execution phase of the experiment, the procedures for single tests should be used.

4-3.4 Precision Error Propagation: Effect of Small Sample Sizes

For single test experimental situations in which all of the precision indices are determined from multiple measurements with some having small samples ($N < 30$), the procedure outlined in Sections 4-3.1 and 4-3.2 and Figure 4.6 must be modified. In such cases, the number of degrees of freedom associated with each precision index must be found and the equivalent degrees of freedom ν_r in the precision index of the result S_r determined. This value of ν_r is then used with the t-distribution to find the value of t such that (tS_r) corresponds to a 95% confidence interval. This procedure is illustrated in the flow diagram of Figure 4.9.

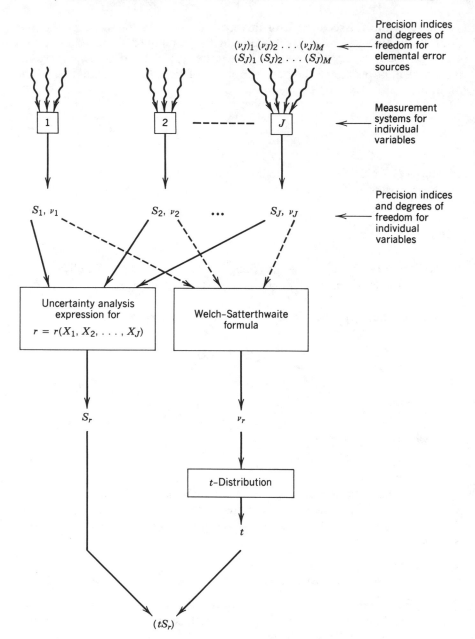

Figure 4.9   Detailed uncertainty analysis—determination of precision limit for experimental result. Result determined from a single test. All precision indices calculated from multiple measurements, but some sample sizes small ($N \leqq 30$).

At this point, it is useful to recall that the precision index can be calculated from a sample of size N as

$$S_{X_i} = \left\{ \frac{1}{N-1} \sum_{k=1}^{N} [(X_i)_k - \bar{X}_i]^2 \right\}^{1/2} \tag{4.34}$$

The associated number of degrees of freedom for use in the t-distribution is equal to $(N-1)$.

When there are elemental error sources that cause the precision error in a variable as shown in Figure 4.9, they must be combined to obtain the precision index for the variable. If for the Jth variable, we are able to identify, isolate, and determine the precision indices by Eq. (4.34) from M significant elemental error sources, we have

$$(S_J)_1, (S_J)_2, \ldots, (S_J)_M$$

These are combined using root-sum-square to obtain

$$S_J = \left[(S_J)_1^2 + (S_J)_2^2 + \cdots + (S_J)_M^2 \right]^{1/2}$$

$$= \left[\sum_{k=1}^{M} (S_J)_k^2 \right]^{1/2} \tag{4.35}$$

The number of degrees of freedom associated with S_J is calculated using the Welch–Satterthwaite formula [1]

$$\nu_J = \frac{\left[\sum_{k=1}^{M} (S_J)_k^2 \right]^2}{\sum_{k=1}^{M} \left[(S_J)_k^4 / (\nu_J)_k \right]} \tag{4.36}$$

where $(\nu_J)_k = (N_J)_k - 1$.

Once the precision index and associated degrees of freedom have been determined for each of the variables X_i in the data reduction equation, the precision index of the result, S_r, is found from the uncertainty analysis expression

$$S_r = \left[\left(\frac{\partial r}{\partial X_1} S_1 \right)^2 + \left(\frac{\partial r}{\partial X_2} S_2 \right)^2 + \cdots + \left(\frac{\partial r}{\partial X_J} S_J \right)^2 \right]^{1/2} \tag{4.37}$$

The number of degrees of freedom, ν_r, associated with S_r is then determined from the Welch–Satterthwaite formula [1] that may be written as

$$
\nu_r = \frac{\left[\displaystyle\sum_{i=1}^{J} (\theta_i S_i)^2\right]^2}{\displaystyle\sum_{i=1}^{J} \left[(\theta_i S_i)^4/\nu_i\right]} = \frac{S_r^4}{\displaystyle\sum_{i=1}^{J} \left[(\theta_i S_i)^4/\nu_i\right]}
\tag{4.38}
$$

or in the alternate form

$$
\nu_r = \frac{\left\{\displaystyle\sum_{i=1}^{J} \left[\theta_i'(S_i/X_i)\right]^2\right\}^2}{\displaystyle\sum_{i=1}^{J} \left[(\theta_i' S_i/X_i)^4/\nu_i\right]} = \frac{(S_r/r)^4}{\displaystyle\sum_{i=1}^{J} \left[(\theta_i' S_i/X_i)^4/\nu_i\right]}
\tag{4.39}
$$

where the θ_is have been previously defined in (4.6) as

$$
\theta_i = \frac{\partial r}{\partial X_i}
\tag{4.6}
$$

and the θ_i' is defined as

$$
\theta_i' = \frac{\partial r/r}{\partial X_i/X_i} = \left(\frac{X_i}{r}\right)\theta_i
\tag{4.40}
$$

The appropriate value of t is then found by entering the table for the t-distribution (Table A.2) with ν_r. The uncertainty in the result is then expressed as

$$
U_{r_{\text{RSS}}} = \left[B_r^2 + (tS_r)^2\right]^{1/2}
\tag{4.41}
$$

for a 95% coverage, or as

$$
U_{r_{\text{ADD}}} = B_r + tS_r
\tag{4.42}
$$

for a 99% coverage.

It should be stressed that the method outlined in this section is used only when a single determination of the result, r, is made and when every one of the precision indices S_i is determined from multiple measurements of its variable X_i and some of the sample sizes have $N \leq 30$. Since in most cases some precision errors will have to be estimated, the general case presented in Sections 4-3.1 and 4-3.2 is normally used.

4-3.5   Determining Precision Limits

In the initial phases of an experimental program before instrumentation is acquired and installed, we make 95% confidence estimates of precision limits ($\pm P$) in much the same manner as estimates of the bias limits are made. At this stage in the experiment, the precision limit associated with the measurement system may be the only precision error source that we consider (as in a zeroth-order analysis). As a general rule of thumb, the precision limit associated with the readability of an analog instrument can be taken as one-half of the least scale division. Of course, judgment must be used in applying this rule. For instance, if an analog instrument has a rather coarse scale division, the precision limit might be less than one-half of the least scale division. For a digital output, the maximum precision limit resulting from the readability (assuming no flicker in the indicated digits) is one-half of the least digit in the output. (See Section 5-2 for a more detailed discussion of errors associated with digital data acquisition.)

Later in the experimental program when we are able to make measurements of a variable under conditions similar to those in the "real" experiment or to make multiple measurements of the variable during the actual experiment, the precision index S can be calculated from a data sample of N points. When such statistical determinations are made, it is very important to keep in mind the time period over which the N data points are acquired.

As an example, consider that 1000 readings are taken of the variable Y while the experimental apparatus is operating at a "constant" set point. The precision index S_Y can then be calculated, and since the number of degrees of freedom is 999 we would suppose that ($2 S_Y$) would be an excellent estimate of the 95% confidence interval. This is true from a strictly statistical standpoint, but we must always be aware of the meaning of S_Y. Stated another way, we must always be aware of the factors that influenced the random variations in the data set of N points. If the data points were taken over a period of 0.1 sec using a high-speed data acquisition system, then the influence of factors that vary with periods of seconds, minutes, hours, or days would not be included in the S_Y calculated.

This idea is illustrated in a very simple manner in Figure 4.10. If Y varies with time and a data set is taken over the time period Δt shown, then neither the average value of Y nor the precision index calculated from the data set will be good estimates of the values appropriate for a time period of 100 times Δt, for example. *Data sets for determining estimates of precision indices should be acquired over a time period that is large relative to the time scales of the factors that have a significant influence on the data and that contribute to the precision errors.*

Barometric pressure varies with a time scale of the order of hours, as does relative humidity. If these factors are uncontrolled but do influence a variable in the experiment, then this should be recognized. If an experiment is sensitive to the vibrations transmitted through the structure of a building, these

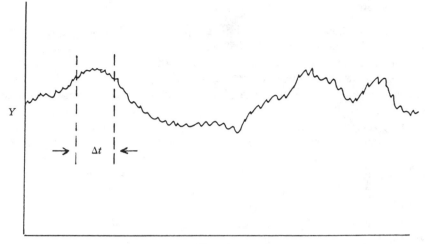

Figure 4.10   Variation of a variable, Y, with time, t, for a "constant" experimental condition.

vibrations may be quite different in morning and afternoon as various pieces of equipment in the building are operated or shut down. There are factors such as these that must be considered for each individual experiment.

4-4   USING DETAILED UNCERTAINTY ANALYSIS

Now that we have covered the basic methodology involved in both general and detailed uncertainty analysis, it is appropriate to consider how these techniques are useful in different phases of an experimental program and how they fit together with the concepts of different orders of replication level that we have discussed previously. An overview of some of the uses of uncertainty analysis in the different phases of an experimental program is given in Figure 4.11.

As we saw in Chapter 3, general uncertainty analysis is used in the planning phase of an experimental program and is useful in ensuring that a given experiment can successfully answer the question of interest. Some decisions in the preliminary design of an experiment can be made based on the results of a general uncertainty analysis.

Once past the planning and preliminary design phases, the effects of bias errors and precision errors are considered separately using the techniques of detailed uncertainty analysis presented in the previous sections of this chapter. This means that estimates of bias limits and precision limits will be made and used in the design phase, then in the construction, debugging, execution, and

Phase of Experiment	Type of U. A.	Uses of U. A.
Planning	General	Choose experiment to answer a question; preliminary design.
Design	Detailed	Choose instrumentation (zeroth order estimates); detailed design (Nth order estimates).
Construction	Detailed	Guide decisions on changes, etc.
Debugging	Detailed	Verify and qualify operation - first order and Nth order comparisons.
Execution	Detailed	Balance checks and monitoring operation of apparatus; choice of test points run.
Data Analysis	Detailed	Guide to choice of analysis techniques.
Reporting	Detailed	Bias limits, precision limits and overall uncertainties reported.

Figure 4.11   Uncertainty analysis in experimentation.

data analysis phases, and finally in the reporting phase of an experiment as shown in Figure 4.11. If one thinks about it, it soon becomes obvious that there will almost always be much more information available with which to make the bias limit and precision limit estimates in the later phases of an experiment than in the earlier phases. This means that estimates of a particular bias limit or precision limit may very well change as an experimental program progresses and we obtain more information—this should be expected and accepted as a fact of life. *Bias limits and precision limits must be estimated using the best information available at the time. Lack of information is no excuse for not doing an uncertainty analysis in the early phases of an experimental program—it is simply a reason why the estimates may not be as good as they will be later in the program.*

The manner in which these estimates are used can differ in timewise and sample-to-sample experiments. Recall that timewise experiments are those in which generally a given entity is tested, either as a function of time or at some steady-state condition in which data are taken over some period of time. Examples would be testing the performance characteristics of a given engine, determining the friction factor for a given pipe over a range of Reynolds numbers, and determining the heat transfer coefficient from a given object immersed in a fluid over a range of flow conditions. Sample-to-sample experiments are those in which generally some characteristic is determined for sample after sample, often with the variability from sample to sample being significant. In this case, sample identity can be viewed as analogous to time in a timewise experiment. Examples would be determining the heating value of a certain type of coal, determining the ultimate strength of a certain alloy, or determining some physical characteristic of a manufactured product for quality control purposes.

In the early stages of the design phase of a program, estimates of the bias limits and precision limits at the zeroth-order replication level are useful in choosing instrumentation and measurement systems. For timewise experiments, this means making the estimates while hypothesizing a totally steady process and environment. For sample-to-sample experiments, it means making the estimates assuming a single, fixed sample. The zeroth-order bias and precision limit estimates indicate the "best case" for a given measurement system in both types of experiment.

When we move beyond this stage in a timewise experiment, we make estimates at the first-order and Nth-order levels. Here we consider all the factors that will influence the bias and precision errors in the experiment. At the first-order replication level, we are interested in the variability of the experimental results for a given experimental apparatus. The descriptor for this variability is P_r, the precision limit of the result. In a timewise experiment, comparison of the estimated P_r and the observed scatter in results from multiple trials at a given set point of the experimental apparatus is useful in the debugging phase. This is discussed in detail with an example in Chapter 6.

In a sample-to-sample experiment, first-order estimates of P_r made before multiple samples are tested are often not very useful, since the variation from sample to sample is usually unknown and is one of the things we discover with the experiment. After multiple samples have been tested, the difference (in a root-sum-square sense) between the calculated P_r from the multiple results (Section 4-3.3) and the zeroth-order precision limit estimate can be used as an estimate of the precision contribution due to sample-to-sample variability. This is illustrated in the lignite heating value determination example in Section 4-5.

In asking questions and making comparisons at the Nth-order replication level, we are interested in the interval within which "the truth" lies. This interval is described by U_r, the overall uncertainty in the result, which is found by combining the first order precision limit P_r and the bias limit B_r using Eq. (4.1)

$$U_r = \left[B_r^2 + P_r^2 \right]^{1/2} \tag{4.1}$$

Comparisons of experimental data with theoretical results or with data from other experiments should be made at the Nth-order replication level. This is discussed in more detail in following chapters.

4-5   USE OF DETAILED UNCERTAINTY ANALYSIS IN A SAMPLE-TO-SAMPLE EXPERIMENT

A mining company has retained our laboratory to determine the heating value of some lignite in which they have an interest. Lignite is a brownish coal that is relatively high in ash and moisture content. The most abundant supplies of lignite in the United States are in the Gulf Coast region and in the Northern

Great Plains region. The company wants to develop a North Dakota lignite deposit into a boiler fuel.

4-5.1 The Problem

Initially the company wants to determine the heating value of a small portion of the deposit of lignite. They plan to deliver to our laboratory sealed samples from this section of the deposit. From these they want answers to the following questions:

   a. How well can we measure the heating value of a single sample?
   b. What is the probable range of heating values for the particular portion of the deposit?

In question (a), we are essentially considering the uncertainty at a zeroth-order replication level as we are concerned only with the uncertainty caused by the bias and precision errors in the instrumentation system. To answer question (a), we will consider the bias and precision errors in the measurement system that combine to give the uncertainty in a single heating value determination.

We know that lignite can have significant variations in the ash and moisture content even within a single deposit. For this reason, we expect that there will be variations in the heating value determinations among the samples because of the physical variability of the material. Therefore, the uncertainty found for question (a) will probably be different from that found for question (b). The uncertainty found for (b) will essentially be at an Nth-order replication level at which variations in the material are taken into account along with the bias and precision errors in the measurement system.

4-5.2 The Measurement System

The heating value will be determined with an oxygen bomb calorimeter (see Figure 4.12). This device is a standard, commercially available system for making this type of determination. A measured sample of the fuel is enclosed in the metal pressure vessel that is then charged with oxygen. This bomb is then placed in a bucket containing 2000 g of water and the fuel is ignited by means of an electrical fuse. The heat release from the burning fuel is absorbed by the metal bomb and the water. By measuring the temperature rise $(T_2 - T_1)$ of the water, the heating value of the fuel can be determined from an energy balance on the calorimeter, which gives the following expression

$$H = \frac{(T_2 - T_1)C - e}{M} \tag{4.43}$$

where H is the heating value (cal/g), T is temperature (F or R), C is the calibrated energy equivalent of the calorimeter (cal/R), e is the energy release from the burned fuse wire (cal), and M is the mass of the fuel sample (g).

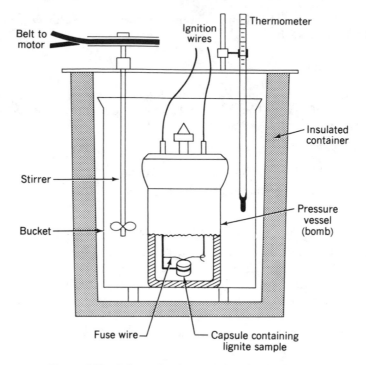

Figure 4.12   Schematic of oxygen bomb calorimeter.

Note that this is the higher heating value as the vapor formed is condensed to liquid.

In the discussion that follows, we will first consider the application of detailed uncertainty analysis at the zeroth-order replication level. Next we will consider the phase of the experiment in which heating values have been determined from multiple samples. These data are then used to calculate the precision limit associated with the measurement system and with the ash and moisture variations in the samples—this corresponds to analysis at the first-order replication level. Finally, we will look at the Nth-order replication level at which the precision errors from the variability of the material and from the measurement system and the bias errors from the measurement system are combined to give us the uncertainty in the heating value for the portion of the lignite deposit being analyzed.

4-5.3 Zeroth-Order Replication Level Analysis

The heating value of each lignite sample will be determined from data obtained with the bomb calorimeter system using the data reduction equation

$$H = \frac{(T_2 - T_1)C - e}{M} \tag{4.43}$$

The calibration "constant" for the system has been previously determined by using a high-purity benzoic acid as the fuel. In this calibration, multiple tests were run and the mean value for C was found to be 1385 cal/R. The precision limit of this mean value was ± 8 cal/R, and the bias limit for the calibration constant was determined to be ± 2 cal/R. Therefore, the uncertainty in the calibration constant is

$$U_C = \left[(8)^2 + (2)^2\right]^{1/2} = \pm 8.2 \text{ cal/R} \qquad (4.44)$$

When C is used in Eq. (4.43), the uncertainty associated with that number will always be ± 8.2 cal/R. Even though this uncertainty was originally obtained from combining precision and bias limits, it becomes "fossilized" into a bias limit for all future calculations when the fixed value of $C = 1385$ cal/R is used.

The temperatures T_1 and T_2 are measured with a thermometer that has 0.05°F as its least scale division. To determine the precision limit for measuring temperatures with this thermometer, different people were asked to read a steady temperature. The precision limit calculated from these readings was ± 0.052°F. Therefore, a single temperature measurement will have this value as the estimate of its precision limit. The thermometer has not been calibrated against a known standard so the bias limit is estimated to be, say, ± 2°F. As will be seen, this bias will not affect the uncertainty in the result because of correlated effects in the T_1 and T_2 measurements.

The fuse wire comes with a scale marked in calories with a least scale division of 1.0 cal. With the scale division, we estimate the precision limit for the measurement of a length of wire to be ± 0.5 cal. The fuse wire correction e used in Eq. (4.43) is the difference between the original wire length and the length of the unburned wire recovered from the bomb after the test is run. Using Eq. (4.23), the precision limit for e is then the square root of the sum of the squares of the precision limits for the two wire lengths or ± 0.7 cal. We assume the bias error for each length measurement would be equal and perfectly correlated, so that the bias in e [from Eq. (4.15)] is zero.

The mass is measured with a digital balance that has a resolution of 0.0001 g. From experience in observing the "jitter" in the digital readout, we estimate that the precision limit for a mass measurement is ± 0.0002 g. Also, the balance has been calibrated with a known mass that has an accuracy of 0.0001 g. The mass M in Eq. (4.43) is similar to the fuse wire length in that it is the difference between two measurements—the mass of the empty sample container and the mass of the container with the lignite sample in it. Therefore the precision limit for M is ± 0.00028 g [using Eq. (4.23) to combine the two mass precision limits] and the bias limit is zero due to correlated bias effects. The zeroth-order estimates of the bias and precision limits for variables are summarized in Table 4.1.

TABLE 4.1   Zeroth-Order Estimates of the Bias and Precision Limits for Variables in Heating Value Determination

Variable	Bias limit	Precision limit
T_1	$\pm 2°F$	$\pm 0.052°F$
T_2	$\pm 2°F$	$\pm 0.052°F$
C	± 8.2 cal/R	0
e	0	± 0.7 cal
M	0	± 0.00028 g

Referring back to the heating value expression,

$$H = \frac{(T_2 - T_1)C - e}{M} \tag{4.43}$$

the partial derivatives needed for the uncertainty analysis are

$$\frac{\partial H}{\partial T_2} = \frac{C}{M} \tag{4.45}$$

$$\frac{\partial H}{\partial T_1} = -\frac{C}{M} \tag{4.46}$$

$$\frac{\partial H}{\partial C} = \frac{T_2 - T_1}{M} \tag{4.47}$$

$$\frac{\partial H}{\partial e} = -\frac{1}{M} \tag{4.48}$$

$$\frac{\partial H}{\partial M} = \frac{-[(T_2 - T_1)C - e]}{M^2} = -\frac{H}{M} \tag{4.49}$$

Because T_1 and T_2 are measured with the same thermometer, the bias limits for these two variables are correlated. The uncertainty analysis expression which must be used for determining B_H is then given by Eq. (4.9), which reduces to the form given by Eq. (4.15) when biases in only two measured variables are correlated. We thus have

$$B_H^2 = \left(\frac{\partial H}{\partial T_2} B_{T_2}\right)^2 + \left(\frac{\partial H}{\partial T_1} B_{T_1}\right)^2 + \left(\frac{\partial H}{\partial C} B_C\right)^2 + \left(\frac{\partial H}{\partial e} B_e\right)^2$$

$$+ \left(\frac{\partial H}{\partial M} B_M\right)^2 + 2\frac{\partial H}{\partial T_2} \frac{\partial H}{\partial T_1}(1) B'_{T_2} B'_{T_1} \tag{4.50}$$

TABLE 4.2   Nominal Values from a Previous Test

Measured variable	Value
T_1	75.25°F
T_2	78.55°F
C	1385 cal/R
e	12 cal
M	1.0043 g

$$H = \frac{(T_2 - T_1)C - e}{M} = 4539 \text{ cal/g}$$

Substituting for the partial derivatives and noting that $B_{T_1} = B_{T_2} = B'_{T_1} = B'_{T_2} = B_T$, we get

$$B_H^2 = \left(\frac{C}{M}\right)^2 B_{T_1}^2 + \left(\frac{C}{M}\right)^2 B_{T_2}^2 + \left(\frac{T_2 - T_1}{M}\right)^2 B_C^2$$

$$+ \left(\frac{1}{M}\right)^2 B_e^2 + \left(\frac{H}{M}\right)^2 B_M^2 - 2\left(\frac{C}{M}\right)^2 B_T^2 \tag{4.51}$$

The terms containing B_T add to zero; therefore, the bias error estimate in the temperature measurements has no effect on the bias limit of the heating value. This occurs because (1) the partial derivatives with respect to T_2 and T_1 are equal in magnitude but opposite in sign [see Eqs. (4.45) and (4.46)], and (2) the entire bias limits estimated for T_2 and T_1 are taken to be perfectly correlated with each other. These two circumstances cause the first, second, and sixth terms on the right-hand side of Eq. (4.50) to add to zero.

We can see from Eq. (4.51) that nominal values of the test variables must be found before a numerical value for B_H can be determined. In this instance, we have available the results from a previous test, so we use those values (Table 4.2) along with the bias limit estimates in Table 4.1. The bias limit in the result then becomes

$$(C) \ (e) \ (M)$$

$$B_H^2 = 726 + 0 + 0 \tag{4.52}$$

or

$$B_H = \pm 27 \text{ cal/g} \tag{4.53}$$

From this calculation, we see that the "fossilized" error in the calibration constant determines the bias error in the heating value.

The precision limit of the result is found using Eq. (4.23), which gives

$$P_H^2 = \left(\frac{\partial H}{\partial T_2} P_{T_2}\right)^2 + \left(\frac{\partial H}{\partial T_1} P_{T_1}\right)^2 + \left(\frac{\partial H}{\partial C} P_C\right)^2$$
$$+ \left(\frac{\partial H}{\partial e} P_e\right)^2 + \left(\frac{\partial H}{\partial M} P_M\right)^2 \tag{4.54}$$

Substituting for the partial derivatives,

$$P_H^2 = \left(\frac{C}{M}\right)^2 P_{T_2}^2 + \left(\frac{C}{M}\right)^2 P_{T_1}^2 + \left(\frac{T_2 - T_1}{M}\right)^2 P_C^2$$
$$+ \left(\frac{1}{M}\right)^2 P_e^2 + \left(\frac{H}{M}\right)^2 P_M^2 \tag{4.55}$$

Using the values given in Tables 4.1 and 4.2, the precision limit becomes

$$\begin{array}{ccccc} (T_2) & (T_1) & (C) & (e) & (M) \\ P_H^2 = 5143 + 5143 + & 0 & + 0.5 + & 1.6 \end{array} \tag{4.56}$$

or

$$P_H = \pm 101 \text{ cal/g} \tag{4.57}$$

Here the temperature precision limit is controlling.

The original question of how well we can measure the heating value of a single sample can now be answered. The uncertainty (with a 95% coverage) in the heating value determination is found using Eq. (4.1) as

$$U_H = \left[B_H^2 + P_H^2\right]^{1/2} \tag{4.1}$$

Using the bias and precision limits previously calculated, the uncertainty becomes

$$U_H = \left[(27)^2 + (101)^2\right]^{1/2} = \pm 105 \text{ cal/g} \tag{4.58}$$

This uncertainty, which represents the zeroth-order replication level estimate for this experiment, is about 2% of the lignite heating value.

4-5.4 First-Order Replication Level Analysis

After the mining company delivered 26 sealed samples to the laboratory, the heating value of each was determined using the bomb calorimeter. These heating value results are summarized in Table 4.3.

TABLE 4.3 Heating Value Results for 26 Lignite Samples

Sample number	Heating value (cal/g)
1	4572
2	4568
3	4547
4	4309
5	4383
6	4354
7	4533
8	4528
9	4383
10	4546
11	4539
12	4501
13	4462
14	4381
15	4642
16	4481
17	4528
18	4547
19	4541
20	4612
21	4358
22	4386
23	4446
24	4539
25	4478
26	4470

Because we have multiple results, we can use the techniques of Section 4-3.3 and directly calculate a precision limit and mean value for H. Using Eq. (4.28), the mean of these heating value results is

$$\overline{H} = 4486 \text{ cal/g} \tag{4.59}$$

From Eq. (4.29), the precision index of the collection of 26 results is

$$\hat{S}_H = 87.2 \text{ cal/g} \tag{4.60}$$

For the 26 data points, the value of t from Table A.2 ($\nu = 25$) is 2.060, and so

$$P_H = \pm t\hat{S}_H = \pm 180 \text{ cal/g} \tag{4.61}$$

The precision limit in Eq. (4.61) represents the range around the mean value at which the heating value determination for another sample from the same portion of the deposit would fall with 95% confidence. This precision limit is greater than that (101 cal/g) found for a single sample, which seems reasonable as it includes contributions to the precision both from the measurement system and from the material composition variation from sample to sample. It is the precision limit appropriate for use in an analysis at the first–order replication level.

If we divide the contributors to the first-order precision limit into two categories—zeroth-order and sample-to-sample material variation—we can write

$$(P_H)^2_{1st} = (P_H)^2_{zeroth} + (P_H)^2_{mat} \tag{4.62}$$

Since the first-order precision limit has been determined as 180 cal/g and the zeroth-order precision limit has been determined as 101 cal/g, the precision limit due to sample-to-sample variation in the lignite composition can be found as

$$(P_H)_{mat} = \left[(P_H)^2_{1st} - (P_H)^2_{zeroth}\right]^{1/2}$$

$$= \left[(180)^2 - (101)^2\right]^{1/2} = \pm 149 \text{ cal/g} \tag{4.63}$$

This estimate is useful because it tells us that the precision limit due to the variability of lignite composition within the portion of the deposit investigated is actually greater than the precision limit associated with our measurement system.

The analysis presented in this section illustrates the primary use of an analysis at the first-order level in a sample-to-sample type experiment. We were able to estimate the variability in the result due to material variations from sample to sample once we had a zeroth-order precision limit estimate and a first-order precision limit calculated directly from multiple results.

4-5.5 Nth-Order Replication Level Analysis

We are now in a position to determine what the probable range of heating values will be for the particular portion of the lignite deposit being considered. This range is given by $\overline{H} \pm U_H$ where $\overline{H}$ is the mean of the results (4486 cal/g) and U_H is the Nth-order estimate of the uncertainty as given by

$$U_H = \left[B_H^2 + P_H^2\right]^{1/2} \tag{4.1}$$

Here, P_H is the first-order estimate of the precision limit found in Eq. (4.61) and B_H is the bias limit found at the zeroth-order replication level in Eq.

(4.53). Thus,

$$U_H = \left[(27)^2 + (180)^2\right]^{1/2} = \pm 182 \text{ cal/g} \tag{4.64}$$

Note that in this particular case, the bias contribution is negligible and the uncertainty is dominated by precision error contributions. Combining $\bar{H}$ and U_H we find that the probable range of heating values will be 4304 to 4668 cal/g.

Now that we have answered the original two questions, the mining company wants us to determine the average heating value of the lignite in the portion of the deposit being analyzed. With the 26 samples that were tested, we found the mean of the results to be 4486 cal/g.

The uncertainty in this average result is, from Eq. (4.31),

$$U_{\bar{H}} = \left[B_{\bar{H}}^2 + P_{\bar{H}}^2\right]^{1/2} \tag{4.31}$$

where

$$P_{\bar{H}} = t S_{\bar{H}} \tag{4.33}$$

and

$$S_{\bar{H}} = \frac{\hat{S}_H}{\sqrt{M}} \tag{4.30}$$

Substituting for $\hat{S}_H$ from Eq. (4.60) we obtain

$$S_{\bar{H}} = \frac{87.2 \text{ cal/g}}{\sqrt{26}} = 17.1 \text{ cal/g} \tag{4.65}$$

with

$$P_{\bar{H}} = (2.060)(17.1 \text{ cal/g}) = 35.2 \text{ cal/g} \tag{4.66}$$

and

$$U_{\bar{H}} = \left[(27)^2 + (35.2)^2\right]^{1/2} = \pm 44 \text{ cal/g} \tag{4.67}$$

where the bias limit found at the zeroth-order replication level in Eq. (4.53) applies here as well. Therefore, the average lignite heating value is, with a 95% coverage, 4486 ± 44 cal/g.

It should be noted that in this experiment, the only significant bias error sources were all present at the zeroth-order replication level. This made the Nth-order bias limit the same as the zeroth-order bias limit. This is not true in general, as often biases arise due to installation, sensor/environment interaction, and other sources not considered at the zeroth level. In general $(B_r)_{N\text{th}} > (B_r)_{\text{zeroth}}$.

4-6   USE OF DETAILED UNCERTAINTY ANALYSIS IN A TIMEWISE EXPERIMENT

The owner of a small manufacturing operation wants to monitor his process by watching, among other things, for changes in the velocity of air exiting an exhaust duct in the system. To specify what magnitude of change in the measured velocity should be viewed as a drift in the operating point of the system, one must know the range of velocity measurements expected if the system is operating at a "steady" set point. The owner wants us to analyze this situation and estimate the expected range of velocity readings at a point in the duct exit plane if the measurements are taken by different employees at different times over a period of weeks using the equipment currently in place. He also wants the expected range if, at some point, he replaces the current measurement system with "new, but identical" equipment. The plant is shut down for scheduled maintenance, so we are unable to actually make measurements before we must give the owner our initial estimates.

4-6.1   The Measurement System

The measurement system consists of a Pitot tube mounted in the exit plane of the duct, an analog differential pressure gauge with a range of 0 to 2 in. of water and a least scale division of 0.05 in. of water, a thermometer with a least scale division of 1°C, and a barometer with a least scale division of 1 mm Hg. The Pitot tube (sometimes called a Pitot-static tube) senses both stagnation pressure and static pressure, and these are fed into the differential pressure gauge. The gauge then indicates the dynamic pressure, Δp, which is the difference between stagnation and static pressures.

The air velocity at the measurement point is given by [4, 5, 6]

$$V = \left(\frac{2 \Delta p}{\rho} \right)^{1/2} \tag{4.68}$$

where ρ is the density of the air and is determined from

$$p = \rho RT \tag{4.69}$$

Substituting into Eq. (4.68) for the density, the data reduction equation becomes

$$V = \left(\frac{2 R T \Delta p}{p} \right)^{1/2} \tag{4.70}$$

4-6.2 Consideration of Error Sources

After considering the questions that must be answered, it is evident that the precision limit P_V will provide the appropriate estimate of the range of scatter expected when the current equipment is used. This first-order replication level estimate must include the effects of inability to exactly reset the desired operating condition as the plant is shut down each weekend and restarted each Monday morning. The degree of unsteadiness in the air velocity when the system is supposedly in "steady" operation should also be considered, as we know that in such a system steady operation is not exactly constant.

The measurement uncertainty U_V will provide the appropriate estimate of the range of scatter expected if "new, but identical" measurement equipment is installed in place of the current equipment. This Nth-order replication level estimate takes into account the fact that the "identical" equipment will not have the same bias errors as the original equipment and the installation of the new probe will not be exactly the same as for the original probe. These then provide sources of variability in addition to those included in P_V.

The owner indicates that nominal values at the normal operating condition can be taken as 300 K and 760 mm Hg for the air temperature and pressure and $\Delta P = 1.5$ in. H_2O for the dynamic pressure indicated by the differential pressure gauge. After conversations with the owner and several employees about the reset variation and unsteadiness observed in the pressure gauge indicator at a "constant" operation condition, we make precision limit estimates of

$$P_{\Delta p, \text{Reset}} = 0.05 \text{ in. } H_2O \qquad (4.71)$$

for the reset contribution and

$$P_{\Delta p, \text{Unsteady}} = 0.025 \text{ in. } H_2O \qquad (4.72)$$

for the gauge output unsteadiness at a "steady" condition.

Examination of the pressure gauge, barometer, and thermometer along with the manufacturer's specification sheets enables estimates to be made of the bias and precision limits inherent in the instruments. These are the zeroth-order replication level estimates, since they are the best we could expect from the instruments without recalibration.

The differential pressure gauge is specified as having an accuracy $\pm 2.5\%$ of full scale. Since full scale is 2 in. H_2O, we estimate the zeroth-order bias limit as

$$B_{\Delta p, 0} = (0.025)(2.0 \text{ in. } H_2O) = 0.050 \text{ in. } H_2O \qquad (4.73)$$

We note that the least division on the scale is 0.05 in. H_2O and, after looking at the scale, decide that one-half the least division is a reasonable estimate of the zeroth-order precision limit

$$P_{\Delta p,0} = 0.025 \text{ in. } H_2O \tag{4.74}$$

The barometer also has an analog scale. The manufacturer specifies an "absolute accuracy of ± 2 mm Hg," which we take to be a reasonable estimate of the bias limit

$$B_{p,0} = 2.0 \text{ mm Hg} \tag{4.75}$$

The least scale division corresponds to 1 mm Hg, but looking at the scale we observe these least divisions are physically fairly large. We decide one-fourth the least division is a reasonable estimate of the precision limit, so

$$P_{p,0} = 0.25 \text{ mm Hg} \tag{4.76}$$

The least scale division on the thermometer is 1°C and there is no information available on its calibration. In the absence of any other information, we assume

$$B_{T,0} = 2°C \tag{4.77}$$

and use one-half the least scale division to estimate the precision limit

$$P_{T,0} = 0.5°C \tag{4.78}$$

If it turns out that the bias limit on the temperature plays a significant role in anything of importance to the owner, we might need to recommend calibration of the thermometer to reduce the bias limit.

The only other error source of potential significance that comes to mind is the Pitot tube installation bias. As discussed in Refs. 4, 5, and 6, errors can arise in the indicated ΔP due to probe misalignment, viscous effects, and other influences. After considering the information in the references, we decide to include a contribution to the bias error in Δp of 1%, or

$$B_{\Delta p, \text{Install}} = (0.01) \Delta p$$

$$= 0.015 \text{ in. } H_2O \tag{4.79}$$

using the nominal value for Δp.

4-6.3   Calculation of Precision and Bias Limits in the Result

Because Eq. (4.70) is in the special power law form, we can write down by inspection

$$\frac{P_V}{V} = \left[\frac{1}{4}\left(\frac{P_T}{T}\right)^2 + \frac{1}{4}\left(\frac{P_{\Delta p}}{\Delta p}\right)^2 + \frac{1}{4}\left(\frac{P_p}{p}\right)^2 \right]^{1/2} \tag{4.80}$$

where P_T, $P_{\Delta p}$, and P_p are the first-order estimates. Because no additional contributors other than those at zeroth order were identified for T and p measurements, P_T and P_p are given by Eqs. (4.78) and (4.76), respectively. The first-order precision limit for Δp is given by

$$P_{\Delta p} = \left[\left(P_{\Delta p, 0} \right)^2 + \left(P_{\Delta p, \text{Reset}} \right)^2 + \left(P_{\Delta p, \text{Unsteady}} \right)^2 \right]^{1/2}$$

$$= 0.075 \text{ in. } H_2O \tag{4.81}$$

We can now substitute into Eq. (4.80) using nominal values for T, p, and Δp and obtain

$$\frac{P_V}{V} = \overset{(T)}{[0.007} \times 10^{-4} + \overset{(\Delta p)}{6.25} \times 10^{-4} + \overset{(p)}{0.0003} \times 10^{-4}]^{1/2} \tag{4.82}$$

or

$$\frac{P_V}{V} = 2.5\% \tag{4.83}$$

We therefore conclude that, using the current equipment as installed, a velocity measurement should fall within a $\pm 2.5\%$ interval about the mean value of previous measurements 95 times out of 100.

To estimate the interval within which velocity measurements will fall if we consider a sample made up of measurements using the current equipment and also measurements using "new, but identical" equipment, we must calculate U_V using

$$\frac{U_V}{V} = \left[\left(\frac{B_V}{V} \right)^2 + \left(\frac{P_V}{V} \right)^2 \right]^{1/2} \tag{4.84}$$

where P_V/V is the first-order precision limit from Eq. (4.83) and B_V/V is given by

$$\frac{B_V}{V} = \left[\frac{1}{4} \left(\frac{B_T}{T} \right)^2 + \frac{1}{4} \left(\frac{B_{\Delta p}}{\Delta p} \right)^2 + \frac{1}{4} \left(\frac{B_p}{p} \right)^2 \right]^{1/2} \tag{4.85}$$

as there are no correlated bias limits. Here, B_p and B_T are given by Eqs. (4.75) and (4.77) as no bias error sources other than those at the zeroth order are considered for the pressure and temperature measurements. $B_{\Delta p}$, however, is given by

$$B_{\Delta p} = \left[\left(B_{\Delta p, 0} \right)^2 + \left(B_{\Delta p, \text{Install}} \right)^2 \right]^{1/2} \tag{4.86}$$

or

$$B_{\Delta p} = 0.052 \text{ in. } H_2O \tag{4.87}$$

after using Eqs. (4.73) and (4.79).

Substituting into Eq. (4.85) and using the nominal values for p, T, and Δp

$$\frac{B_V}{V} = \overset{(T)}{[0.111} \times 10^{-4} + \overset{(\Delta p)}{3.004} \times 10^{-4} + \overset{(p)}{0.017} \times 10^{-4}]^{1/2} \quad (4.88)$$

or

$$\frac{B_V}{V} = 1.8\% \quad (4.89)$$

We can now substitute into Eq. (4.84) and find

$$\frac{U_V}{V} = 3.1\% \quad (4.90)$$

We conclude, then, that the width of the 95% coverage interval within which we would expect a velocity measurement to fall increases from $\pm 2.5\%$ to $\pm 3.1\%$ when we consider the possibility of replacing the measurement system with a "new, but identical" one.

REFERENCES

1. *Measurement Uncertainty*, ANSI/ASME PTC 19.1-1985 Part 1, 1986.
2. Moffat, R. J., "Describing the Uncertainties in Experimental Results," *Experimental Thermal and Fluid Science*, Vol 1, Jan. 1988, pp 3–17.
3. *Tables of Thermal Properties of Gases*, U.S. National Bureau of Standards Circular 564, 1955.
4. Dally, J. W., Riley, W. F., and McConnell, K. G., *Instrumentation for Engineering Measurements*, John Wiley, New York, 1984.
5. Doebelin, E. O., *Measurement Systems Application and Design*, 3rd ed., McGraw-Hill, New York, 1983.
6. Holman, J. P., *Experimental Methods for Engineers*, 4th ed., McGraw-Hill, New York, 1984.

CHAPTER 4

Problems

4.1   The elemental bias limits affecting readings from a certain pressure transducer have been estimated as 3 psi from the calibration standard, 1 psi from using a linear curvefit equation to represent the calibration data, and 5 psi from installation effects. What is the appropriate estimate for the bias limit of a pressure measurement made using this transducer?

4.2   The elemental bias limits affecting the measurement of temperature with a particular thermistor probe are estimated as 0.2°C from the calibration standard, 0.35°C due to temperature nonuniformity of the "constant temperature" bath used during calibration, 0.5°C from use of a curvefit equation to represent the calibration data, and 0.2°C from installation effects. What is an appropriate estimate for the bias limit of a temperature measurement made using this probe?

4.3 In a testing program to investigate the effectiveness of different lumber drying procedures, a load of lumber is weighed on Scale Y, taken to a nearby kiln for drying, and then is taken to Scale Z, where it is weighed. The amount of moisture lost during the drying operation is determined by subtracting the second weight from the first. The only biases of significance are the calibration bias errors associated with the two scales. What is a reasonable estimate for the bias limit of the moisture determination if

$$B_{cal, Y} = 150 \text{ lbf}$$

$$B_{cal, Z} = 175 \text{ lbf?}$$

4.4 For the situation in Problem 4.3, what would be the effect on the bias limit of the moisture determination if the two scales were calibrated using the same truck as a reference standard, but the weight of the truck is only known ± 500 lbf?

4.5 The rate at which energy is removed from an air-cooled system is being determined by measuring the air inlet and outlet temperatures (T_i and T_0) and the air mass flow rate and using

$$\dot{E} = \dot{m} c_p (T_0 - T_i)$$

where c_p is the specific heat of air at constant pressure and is evaluated at an average temperature. For the operating condition of interest, nominal inlet and outlet temperatures of about 300°K and 320°K are expected. A fossilized bias limit for c_p has been estimated at 1%, and the manufacturer of the mass flow meter guarantees "an absolute accuracy of 0.5% of reading." The contributions to the bias limits for the temperature measurements include a 0.5°K and 0.7°K nonuniformity effect at the inlet and outlet, respectively, and (uncorrelated) calibration bias limits of 1.0°K for both temperature probes. Estimate the bias limit for the result, $\dot{E}$.

4.6 For the situation in Problem 4.5, could there be any improvement by calibrating the two temperature probes against a thermometer with a bias limit of 2.0°K? What would be the new bias limit in $\dot{E}$ be?

4.7 Consider the experimental data for the specific heat of air as presented in Figure 4.4. What would be a reasonable estimate for the fossilized bias limit associated with use of tabulated values of c_p in the 250 to

300°K range? In the 450 to 550°K range? What assumptions do you make in arriving at these estimates?

4.8 Consider the experimental data for the thermal conductivity of air as presented in Figure 4.5. What would be a reasonable estimate for the fossilized bias limit associated with use of tabulated values of k in the temperature range close to 300°K? In the range close to 500°K? What assumptions do you make in arriving at these estimates?

4.9 For the situation in Problem 4.5, replication of the test over a long time period indicates an inherent unsteadiness that results in 95% confidence precision limits of 0.8% for $\dot{m}$ and 0.8°K for each of the temperatures. What is the precision limit for the result, $\dot{E}$? What is the 95% coverage estimate for the overall uncertainty in $\dot{E}$? How does taking advantage of the correlated bias effect (Problem 4.6) change the overall uncertainty in $\dot{E}$?

4.10 Determine the 95% coverage estimate of the uncertainty in the mean value of the ultimate strength of an aluminum alloy determined by taking the average of six experimental results. The 95% coverage bias limit is estimated as 215 psi, and the precision index, S, of the sample of six results is 178 psi.

5

ADDITIONAL CONSIDERATIONS IN EXPERIMENTAL DESIGN

In the previous chapters we have developed the methodology necessary to use uncertainty analysis in the planning and design phases of an experiment. We have defined bias and precision limits, and we have discussed the process of determining estimates of these limits and of combining them to calculate the uncertainty in the result. The example of the lignite heating value determination given in Chapter 4 illustrated the key points of detailed uncertainty analysis.

In this chapter we will present some additional topics that should be considered in the design phase of an experiment. The first of these will be the concept of accounting for transducer installation bias errors either by modifying the data reduction equation or by including these bias errors in the determination of the bias limit of the result. This topic was introduced in Section 4-2.5, and it will be presented in more detail in this chapter through the use of examples.

The next topic will be errors that are inherent with digital data acquisition. Since the use of computer-based data acquisition systems is becoming standard practice, one should understand the errors associated with the analog-to-digital signal conversion process.

The last topic covered in this chapter will be the dynamic response of instrument systems and the errors associated with that response. Transducer response to changing test conditions will be discussed along with the measurement of periodic signals.

5-1 VARIABLE BUT DETERMINISTIC BIAS ERRORS

In some experimental situations, there are significant bias errors that displace the experimental result in a specific direction away from the true result. These asymmetrical bias errors may vary as the experimental result varies, but they are fixed for a given value of the result. Moffat [1] calls this type of error " variable but deterministic."

Transducer installation bias errors discussed earlier are usually of this asymmetrical type. They tend to cause the transducer output to be consistently either higher or lower than the quantity that is being measured. These asymmetrical bias errors must be estimated and included in the determination of the bias limit of the result. This approach is illustrated in the examples below.

In many cases an analytical expression can be formulated for these asymmetrical bias errors, and this expression can be included in the data reduction equation. This technique will usually allow us to reduce the overall uncertainty in the experimental result. Instead of the original asymmetrical bias error, the symmetrical bias and precision limits in the new terms of the data reduction expression become factors in determining the uncertainty. This concept is illustrated in the following example.

5-1.1 Example: Biases in a Gas Temperature Measurement System

A thermocouple probe with a 1/8-in.-diameter stainless-steel sheath is inserted into the exhaust of a 50-horsepower diesel engine as shown in Figure 5.1. The exhaust pipe is a cylinder with a 4-in. diameter. We wish to determine the exhaust gas temperature.

Before conducting an analysis of the measurement system for determining the gas temperature T_g, we must first estimate the conditions in the diesel exhaust. A standard textbook on internal combustion engines [2] can provide the required information. For a diesel engine operating at full throttle and loaded to 50 horsepower, the fuel flowrate would be about 25 lbm/hr and the air flowrate would be about 375 lbm/hr. The exhaust temperature would be in the neighborhood of 1000°F (1460°R) at the exhaust manifold. Assuming that this temperature measurement is to be made in the pipe near the exhaust manifold, then the gas temperature would be about 1000°F and the pipe wall temperature would be lower, maybe 750°F.

The measurement system for this temperature determination is a standard stainless-steel-sheathed thermocouple connected to a digital thermometer. The thermocouple bead is located at the end of the probe in the center of the pipe. For this example we will assume that this system has not been calibrated, but information is available about the accuracy of these devices. The manufacturer's accuracy specification for the thermocouple is " $\pm 2.2°C$ or $\pm 0.75\%$" of the temperature, whichever is greater [3]. The percentage in this case refers to the temperature in °C rather than an absolute temperature as confirmed in a

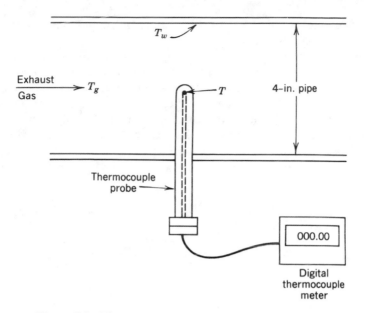

Figure 5.1 Thermocouple probe in engine exhaust pipe.

telephone conversation with the manufacturer. Converting this specification to °F gives an accuracy of $\pm 4.0°F$ or $\pm 0.42\%$ of the temperature in °F minus 32°F [$\pm 0.0042(T - 32°F)$], whichever is greater. The manufacturer's accuracy specification for the digital thermometer is $\pm 2°F$.

We will take these specifications to represent bias errors in the temperature determination. Therefore for this example, the bias error due to the thermocouple probe would be

$$B_{\text{probe}} = \pm(0.0042)(1000°F - 32°F) = \pm 4.1°F \qquad (5.1)$$

Combining the probe and thermometer errors gives us a bias error estimate for the gas temperature measurement of

$$B_{T_g} = \left[(4.1)^2 + (2)^2\right]^{1/2}$$

$$B_{T_g} = \pm 4.6°F \qquad (5.2)$$

Neglecting precision errors, this might be an initial estimate of the range for a 95% coverage of the gas temperature. However, in this case there are other physical processes occurring that will significantly affect the measurement.

Other bias errors influence the temperature determination. There will be some variation in the gas temperature around the pipe center, but for this example we are considering the gas temperature around the tip of the thermocouple probe; therefore, we will consider any temperature variations in this region to be negligible.

A significant bias error will occur in the temperature determination because of the radiation heat loss from the probe to the pipe wall. With the hot gas and the cooler pipe wall, the thermocouple probe will be at a temperature between these two limits. For this example, the thermocouple temperature, T, could be as much as 50 to 70°F below the gas temperature. This error would be an asymmetrical bias since T would always be lower than T_g by this amount for a given operating condition.

Taking all of these bias estimates into account, the bias for the temperature determination would be

$$B_{T_g}^+ = \left[B_{\text{probe}}^2 + B_{\text{meter}}^2 + B_{\text{radiation}}^2 \right]^{1/2}$$

$$B_{T_g}^+ = \left[(4.1)^2 + (2)^2 + (60)^2 \right]^{1/2}$$

$$B_{T_g}^+ = 60°F \tag{5.3}$$

and

$$B_{T_g}^- = \left[B_{\text{probe}}^2 + B_{\text{meter}}^2 \right]^{1/2}$$

$$B_{T_g}^- = \left[(4.1)^2 + (2)^2 \right]^{1/2}$$

$$B_{T_g}^- = 4.6°F \tag{5.4}$$

Because for this case precision errors in the measurement system will be much less than the bias errors, these bias limits also represent the uncertainty interval for a 95% coverage of the gas temperature so that

$$U_{T_g}^+ = 60°F \tag{5.5}$$

and

$$U_{T_g}^+ = 4.6°F \tag{5.6}$$

In the range around 1000°F (1460°R), the reading from the thermometer combined with these uncertainty values would bound the true value of the gas temperature. This uncertainty interval is much larger than our initial estimate of ±4.6°F. This example illustrates the consequences of neglecting a significant physical bias when making a measurement.

Depending on the application for this temperature determination, the above error estimates may or may not be acceptable. The significant bias error in this case is of course the radiation correction. This example is a case in which the physical principles associated with the process, the radiation heat loss, can be incorporated into the data reduction equation to possibly reduce the overall error in the final determination.

Consider an energy balance on the thermocouple probe where

$$\text{Energy in} = \text{Energy out} \tag{5.7}$$

or

$$\text{Convection to the probe} = \text{Radiation from the probe}$$
$$+ \text{Conduction out of the probe} \tag{5.8}$$

The conduction term is much smaller than the radiation and will be neglected. Therefore using expressions for convection and radiation [4], the energy balance becomes

$$h\left(T_g - T\right) = \epsilon\sigma\left(T^4 - T_w^4\right) \tag{5.9}$$

where h is the convective heat transfer coefficient, ϵ is the emissivity of the probe surface, and σ is the Stefan–Boltzmann constant (0.1714×10^{-8} Btu/hr-ft^2-R^4). Note that absolute temperatures in °R are used. Solving this expression for T_g yields the data reduction equation

$$T_g = \frac{\epsilon\sigma}{h}\left(T^4 - T_w^4\right) + T \tag{5.10}$$

Uncertainty analysis can now be used to estimate the error in the gas temperature determination from the errors associated with T, T_w, ϵ, and h. We will still make the assumption that the biases are much greater than the precision errors so that $B_{T_g} = U_{T_g}$. The uncertainty expression for this example is

$$B_{T_g} = \left[\left(\frac{\partial T_g}{\partial \epsilon}B_\epsilon\right)^2 + \left(\frac{\partial T_g}{\partial h}B_h\right)^2 + \left(\frac{\partial T_g}{\partial T}B_T\right)^2\right.$$
$$\left. + \left(\frac{\partial T_g}{\partial T_w}B_{T_w}\right)^2\right]^{1/2} \tag{5.11}$$

where

$$\frac{\partial T_g}{\partial \epsilon} = \frac{\sigma}{h}\left(T^4 - T_w^4\right) \tag{5.12}$$

$$\frac{\partial T_g}{\partial h} = \frac{-\epsilon\sigma}{h^2}\left(T^4 - T_w^4\right) \tag{5.13}$$

$$\frac{\partial T_g}{\partial T} = \frac{4\epsilon\sigma}{h}T^3 + 1 \tag{5.14}$$

and

$$\frac{\partial T_g}{\partial T_w} = \frac{-4\epsilon\sigma}{h}T_w^3 \tag{5.15}$$

Values for the emissivity and the convective heat transfer coefficient are required. These can be obtained from a standard heat transfer textbook [4]. Using a heat transfer correlation for crossflow over a cylinder, a value for h is determined to be 47 Btu/hr-ft²-R. It is known that these heat transfer correlations are accurate to about $\pm 25\%$; therefore, for h the bias error, B_h, is about ± 12 Btu/hr-ft²-R.

The emissivity of a metal surface is very dependent on the surface condition. For the stainless-steel (Type 304) thermocouple probe, the emissivity can vary from 0.36 to 0.73 [5]. Therefore for our initial analysis, we will take ϵ as 0.55 with a bias error of ± 0.19.

The pipe wall temperature has been estimated to be about 750°F (1210°R), but this value could be in error by as much as ± 100°R. This estimate gives us a value for B_{T_w}. The bias error in the temperature determination, T, is the combination of the probe and meter biases so that

$$B_T = \left[(4.1)^2 + (2)^2\right]^{1/2}$$

$$B_T = \pm 4.6\text{°F} \tag{5.16}$$

In summary, the nominal values and error estimates for the quantities used to determine the gas temperature are

$$\epsilon = 0.55 \pm 0.19 \tag{5.17}$$

$$h = 47 \pm 12 \text{ Btu/hr-ft}^2\text{-R} \tag{5.18}$$

$$T = 1460 \pm 4.6\text{°R} \tag{5.19}$$

$$T_w = 1210 \pm 100\text{°R} \tag{5.20}$$

Using the data reduction equation [Eq. (5.10)] and the uncertainty expression [Eq. (5.11)], the gas temperature would be

$$T_g = 1508 \pm 26\text{°R} \tag{5.21}$$

This value is an improvement over our original calculation in which the radiation effect was estimated. But it still might not be good enough depending on the application.

Let us examine the calculation of the bias error in T_g. Substituting the nominal values for ϵ, h, T, and T_w along with their bias estimates into the uncertainty expression yields

$$B_{T_g} = \left[\overset{(\epsilon)}{272} + \overset{(h)}{149} + \overset{(T)}{33} + \overset{(T_w)}{199} \right]^{1/2} \tag{5.22}$$

The uncertainty in the emissivity is the largest contributor to the bias in T_g. A closer examination of the available data for ϵ [5] shows that for Type 304 stainless steel exposed to a temperature near 1000°F for over 40 hr the emissivity varies from 0.62 to 0.73. Therefore after this thermocouple probe has been in service for some time, the value of ϵ would be about 0.68 ± 0.06. These revised values yield the following estimates:

$$B_{T_g} = [\overset{(\epsilon)}{27} + \overset{(h)}{227} + \overset{(T)}{36} + \overset{(T_w)}{304}]^{1/2}$$

$$B_{T_g} = \pm 24°R \tag{5.23}$$

and

$$T_g = 1519 \pm 24°R \tag{5.24}$$

The contribution of the error in ϵ to the total bias error has been significantly reduced, but the increase in the nominal value of ϵ has increased the other terms in the uncertainty expression. The overall uncertainty in the T_g determination is essentially unchanged. But if the thermocouple probe has been in extended service, this estimate of uncertainty is more appropriate.

Considering the effect of the pipe wall temperature on the total error, the determination can be improved by measuring T_w with a surface temperature thermocouple probe and a separate digital thermometer. With such a system for measuring T_w, the bias error for the probe would be

$$B_{\text{probe}} = \pm(0.0042)(750°F - 32°F) = \pm 3°F \tag{5.25}$$

Because this value is less than the minimum probe bias of $\pm 4°F$ or $\pm 4°R$, this latter value will be used. With an estimated bias error of $\pm 4°R$ for the installation of the surface probe and a bias error $\pm 2°R$ for the digital thermometer, the bias in T_w would be

$$B_{T_w} = [(4)^2 + (4)^2 + (2)^2]^{1/2}$$

$$B_{T_w} = \pm 6°R \tag{5.26}$$

These improvements in the overall measurement system yield

$$B_{T_g} = [\overset{(\epsilon)}{27} + \overset{(h)}{227} + \overset{(T)}{36} + \overset{(T_w)}{1}]^{1/2}$$

$$B_{T_g} = \pm 17°R \tag{5.27}$$

and

$$T_g = 1519 \pm 17°R \tag{5.28}$$

The convective heat transfer coefficient bias of $\pm 25\%$ cannot be improved on. The effect of this term by itself gives a bias in T_g of $\pm 15°R$. Therefore, the only additional improvements in the gas temperature determination would come from calibration of the thermocouple probe and digital thermometer system. With calibration, B_T could be reduced to less than $\pm 1°R$ yielding an uncertainty in T_g of

$$T_g = 1519 \pm 16°R \tag{5.29}$$

The above example illustrates the sequence of logic used in designing the gas temperature measurement system. The specific biases used and the associated uncertainty for a 95% coverage of the gas temperature will depend on the required application of the temperature.

5-1.2   Example: Determination of the Velocity Profile in a Pipe

This example is based on an undergraduate engineering laboratory experiment conducted by the authors. The intent of this experiment was to illustrate to the students the concepts of replication and precision limits. In this experiment, the students were instructed to estimate the precision limits associated with determining the velocity profile in a pipe flow for a particular centerline air velocity setting. Estimates were to be made at the first-order level of replication and to take into account the errors in resetting the apparatus to the desired operating condition.

The test setup is shown in Figure 5.2. A variable speed fan was used to establish an air flow through a 1.75-in.-i.d. pipe of length sufficient so that fully developed flow was expected at the pipe exit plane. A Pitot tube was mounted on a traverse so that measurements could be made at different radial positions in a given cross section. A Pitot tube is a probe that has a stagnation pressure tap at its tip and static pressure taps around the circumference. These Pitot tube outputs were connected to an analog differential pressure gauge with a range of 0 to 2 in. of water and a least scale division of 0.05 in. of water. Ambient temperature and pressure were measured with a thermometer and a barometer.

To obtain data for a velocity profile, a student adjusted the fan speed until the differential pressure reading was 1.80 in. of water with the Pitot tube at the pipe centerline. The student then traversed the probe to each of nine radial positions—0.2, 0.4, 0.6, and 0.7 in. on each side of the centerline and at the centerline itself. The differential pressure reading for each position was recorded, and the fan was then turned off. Another student then repeated this procedure. In all, 46 students participated in the experiment.

The data reduction equation for the determination of velocity is

$$V = \left[\frac{2}{\rho} \Delta P \right]^{1/2} \tag{5.30}$$

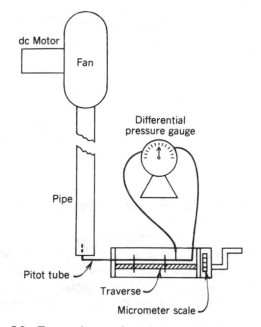

Figure 5.2   Test equipment for velocity profile determination.

where ΔP is the differential pressure reading from the Pitot tube outputs (stagnation pressure minus static pressure) and ρ is the density of the air. The density of the air was determined from the ideal gas relationship as

$$\rho = \frac{P}{RT} \tag{5.31}$$

where P is the atmospheric pressure, R is the gas constant for air, and T is the absolute temperature.

Since the intent of this experiment was for the students to observe the precision errors resulting from setting the flow conditions and then reading the analog pressure gauge, the measurement system bias errors were not initially considered. The density was determined once by measuring the atmospheric pressure and the air temperature. The precision and bias errors associated with this determination then become "fossilized" into a bias error when the density is used in Eq. (5.30). Therefore, the only precision errors propagated into the experimental result, V, come from the precision errors in the Pitot tube pressure difference measurements.

For each of the radial positions, there were 46 determinations of the experimental result, V. These were averaged to give a mean result, $\overline{V}$, at each position. These are given in Table 5.1 along with the precision indices, $\hat{S}_V$, and precision limits, $t\hat{S}_V$. These precision limits define the range within which

TABLE 5.1 Velocity Profile Determination Results for Air Flow in a Pipe

Radial Position (in)	−0.7	−0.6	−0.4	−0.2	0.0	+0.2	+0.4	+0.6	+0.7
Mean velocity, $\overline{V}$ (ft/sec)	78.6	85.1	90.9	91.6	90.3	88.3	85.3	79.8	75.6
Precision index, $\hat{S}_V$ (ft/sec)	1.08	1.05	0.99	0.97	0.90	0.99	0.90	0.94	0.69
Precision limit, $t\hat{S}_V$ (ft/sec)	2.16	2.10	1.98	1.94	1.80	1.98	1.80	1.88	1.38

additional velocity profile determinations will fall with 95% confidence when the same equipment is used.

The velocity profile results are plotted in Figure 5.3. The average velocity determination at each radial position is shown along with the corresponding precision limit. These precision limits represent first-order replication level estimates and give the range of random errors associated with setting a particular flow condition and using the measurement system to determine the velocity.

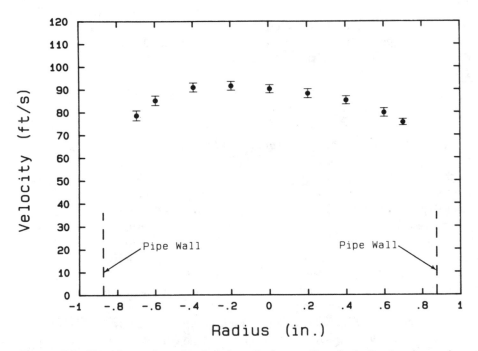

Figure 5.3 Experimental results for the velocity profile of air flowing in a pipe. Precision limits are shown, with no biases considered.

Note that the velocity profile shown in Figure 5.3 is not symmetrical around the pipe centerline as might be expected. The probable cause of this skewed result is the blockage effect of the Pitot tube in the air flow. This effect is another example of a transducer installation bias error, where in this case the presence of the transducer affects the quantity being measured in a nonnegligible manner. Even though this bias does not influence the replication of the velocity profile determination, which was the point of the undergraduate laboratory, we will formulate an approximate analysis to estimate the blockage bias and then will determine overall uncertainties in the measured velocities.

The Pitot tube reduces the flow area as it is traversed across the pipe. This blockage effect causes the flow to accelerate around the tube resulting in a decrease in the static pressure in the vicinity of the static pressure taps. The resulting ΔP reading is then too large causing the velocity determination to be high. This effect is especially important when using a Pitot tube in a small diameter pipe such as the one in this example [6].

The true flow patterns around the Pitot tube are difficult to determine, but we can use the conservation of mass principle to estimate the blockage bias. Assuming that the air mass flow rate is the same for the blocked and unblocked cases, the following relationship will apply:

$$\rho A_{tot} V_t = \rho A V \qquad (5.32)$$

where ρ is the air density, A_{tot} is the total cross-sectional area of the pipe, V_t is the velocity that would be determined if the Pitot tube were not disturbing the flow field, A is the blocked flow area, and V is the velocity determined from the Pitot tube pressure differential.

The measured quantity is the pressure differential instead of velocity. Therefore, by substituting Eq. (5.30) into Eq. (5.32) and rearranging the terms we have

$$\Delta P_t = \left(\frac{A}{A_{tot}} \right)^2 \Delta P \qquad (5.33)$$

where ΔP_t is the Pitot tube pressure differential that would be measured if the tube were not disturbing the flow. The bias limits for ΔP due to the tube blockage can now be estimated as

$$\left(B_{\Delta P}^+ \right)_{blockage} = 0 \qquad (5.34)$$

$$\left(B_{\Delta P}^- \right)_{blockage} = \Delta P - \Delta P_t$$

$$= \Delta P \left[1 - \left(\frac{A}{A_{tot}} \right)^2 \right] \qquad (5.35)$$

The blocked flow area is dependent on the radial position of the Pitot tube as shown in Figure 5.4. The tube is one-eighth in. in diameter and appears as a

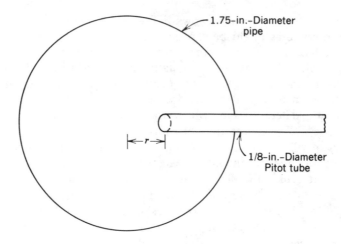

Figure 5.4   Blockage effect of Pitot tube in the pipe.

rectangular cross-sectional obstruction to the flow. The radial position is defined as the distance from the pipe centerline to the center of the Pitot tube opening. Therefore, the blocked area can be expressed as

$$A = A_{\text{tot}} - \left[(0.875 - r)0.125 + \frac{1}{2}\pi \left(\frac{0.125}{2} \right)^2 \right] \tag{5.36}$$

where r is the radial position of the probe mouth in inches. The average measured pressure differential at each radial position is given in Table 5.2 along with the area ratio and the negative blockage bias limit for ΔP calculated using Eq. (5.35).

The blockage bias limits for ΔP must be combined with the pressure gauge bias limit to obtain the overall bias limits for ΔP. The pressure gauge was calibrated against a research micromanometer, and the estimated bias limit for the calibrated gauge is $(B_{\Delta P})_{\text{cal}} = \pm 0.005$ in. of water. Therefore, the bias limits for ΔP are

$$B_{\Delta P}^+ = \left[(B_{\Delta P}^+)_{\text{blockage}}^2 + (B_{\Delta P})_{\text{cal}}^2 \right]^{1/2}$$

$$= 0.005 \text{ in. of water} \tag{5.37}$$

and

$$B_{\Delta P}^- = \left[(B_{\Delta P}^-)_{\text{blockage}}^2 + (B_{\Delta P})_{\text{cal}}^2 \right]^{1/2} \tag{5.38}$$

The values for $B_{\Delta P}^-$ are given in Table 5.2.

TABLE 5.2   Bias Limits and Uncertainty Values for Velocity Profile Determination

Radial Position (in)	−0.7	−0.6	−0.4	−0.2	0.0	+0.2	+0.4	+0.6	+0.7
Average measured pressure differential, ΔP (in. water)	1.36	1.60	1.82	1.85	1.80	1.72	1.60	1.40	1.26
Blockage area ratio, A/A_{tot}	0.916	0.921	0.931	0.942	0.952	0.962	0.973	0.983	0.988
Negative blockage bias limit for ΔP, $(B_{\Delta P}^-)_{blockage}$ (in. water)	0.220	0.243	0.242	0.210	0.168	0.127	0.0862	0.0469	0.0291
Negative bias limit for ΔP, $B_{\Delta P}^-$ (in. of water)	0.220	0.243	0.242	0.210	0.168	0.127	0.0863	0.0472	0.0295
Positive bias limit for velocity, B_V^+ (ft/sec)	0.54	0.58	0.62	0.62	0.61	0.60	0.58	0.55	0.53
Negative bias limit for velocity, B_V^- (ft/sec)	6.38	6.49	6.07	5.23	4.26	3.31	2.37	1.45	1.02
Positive uncertainty in velocity, U_V^+ (ft/sec)	2.23	2.18	2.07	2.04	1.90	2.07	1.89	1.96	1.48
Negative uncertainty in velocity, U_V^- (ft/sec)	6.74	6.82	6.38	5.58	4.62	3.86	2.98	2.37	1.72

The bias limit for the result, B_V, is determined by applying the uncertainty expression, Eq. (4.8), to Eq. (5.30)

$$\left(\frac{B_V}{V} \right) = \frac{1}{2} \left[\left(\frac{B_{\Delta P}}{\Delta P} \right)^2 + \left(\frac{B_\rho}{\rho} \right)^2 \right]^{1/2} \tag{5.39}$$

The bias limit for the density, B_ρ, is the "fossilized" bias or total uncertainty in the single measurement. This value was found to be ± 0.001 lbm/ft^3 with a density of 0.075 lbm/ft^3. By substituting Eqs. (5.37) and (5.38) into Eq. (5.39), the asymmetrical bias limits for the velocity can be determined. These values are given in Table 5.2.

In this experiment, the result is the velocity profile of the air flow in the pipe that exists when a particular flow condition is set (ΔP of 1.8 in. of water at the pipe centerline). The random errors associated with setting the flow condition and using the measurement system were determined previously and are given as the precision limits in Table 5.1. When these are combined in an

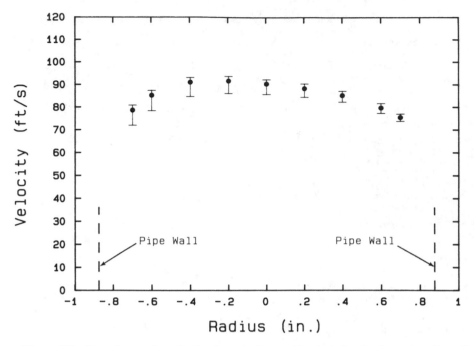

Figure 5.5 / Experimental results for the velocity profile of air flowing in a pipe. Overall uncertainties are shown, with asymmetry of the intervals caused by the blockage bias.

RSS manner with the bias limits for the velocity, the uncertainty for V can be found. These values are given in Table 5.2.

The velocity profile showing the uncertainty at each radial location is presented in Figure 5.5. We can be 95% confident that when the flow conditions are set, the true velocity profile will fall within this range.

This example illustrates another case of variable but deterministic bias errors. In this instance we were able to use a simple approximate analysis to obtain an estimate for the blockage bias. In any experiment, possible transducer–system interactions should be identified and corrected if possible. Sometimes the use of analytical expressions will be sufficient as illustrated in the examples presented here. In other situations, it may be necessary to redesign the test to minimize system–sensor interactions, such as using a shielded thermocouple for the example given in Section 5-1.1 and using a smaller diameter Pitot tube for the example given above.

5-2 DIGITAL DATA ACQUISITION

The use of microcomputers in data acquisition has become common with the availability of relatively inexpensive analog-to-digital (A/D) converters. These devices convert the analog signal from the transducer to a digital signal that can be recorded by the computer and used in a data reduction program.

In this section we will first consider the digitizing error resulting from the A/D conversion process. Then we will discuss the bias and precision errors associated with digital instruments.

5-2.1   Analog-to-Digital Conversion

As described in texts on engineering measurement systems [6, 7], the two main types of A/D converters are the shift-programmed, successive-approximation type and the dual-slope, integrating type. The former is better for high-speed applications and the latter is usually cheaper and useful for lower speed applications. We will not attempt to describe the operation of these systems, but will depend on the references [6, 7] to provide these details. Our purpose here is to consider the error that the digitizing process introduces into a measurement.

Consider the operation of a digital data acquisition system which includes the basic A/D converter chip. Digitizing systems are available from various manufacturers that contain the necessary amplifiers to scale the input analog signal to match the required input range for the A/D converter (usually ± 5 V). Therefore, the calibrated input voltage range for an A/D converter system will vary depending on the signal conditioning used in the system. Digital data acquisition systems usually have options that allow for $\pm$ millivolt ranges or for $\pm$ volt ranges.

In the A/D converter system, a calibrated voltage range is divided into discrete parts in which the resolution is dependent on the number of bits used by the unit. An 8-bit A/D converter will divide the voltage range into 2^8 parts or 256, whereas a 12-bit system will improve the resolution to 2^{12} or 4096 parts. The bias error associated with the digitizing process will then depend on this resolution.

The possible error resulting from the digitizing process can be as large as 1/2 LSB (least significant bit). The resolution of an LSB is equal to the full-scale voltage range divided by 2^N where N is the number of bits used in the A/D converter [7]. The bias error associated with the digitizing process is then taken as $\pm 1/2$ LSB. This bias will be in addition to the other biases inherent in the instrumentation system (such as gain, zero, and nonlinearity errors).

Consider an example of an A/D converter system that has been calibrated for a range of 0 to 10 V full scale. With an 8-bit system, the bias error due to digitizing an input analog signal would be $\pm 1/2(10$ V$/256)$ or ± 0.020 V. For a 12-bit system the digitizing bias error would be reduced to $\pm 1/2(10$ V$/4096)$ or ± 0.0012 V.

It should be stressed that these errors are based on the full-scale calibration of the A/D converter. Care must be exercised in using the appropriately calibrated voltage range for the transducer being read. Consider for example a given transducer that has an output of 100 mV. If a 10 V A/D converter system were used to measure this signal, then the digitizing bias error for an 8-bit system would be $\pm 20\%$. For a 12-bit system this error would be reduced to 1.2%. However, if a 100-mV calibration range were used for the A/D converter instead of 10 V, the digitizing bias errors would be very small for either an 8-bit or a 12-bit system. Both the voltage calibration range and number of bits used by the system must be considered when using digital data acquisition.

5-2.2 Digital Instrument Errors

In a digital instrument, the input analog signal is assigned to a specific digital location. This process can introduce a digitizing bias error as discussed in the previous section and can also influence the precision error of the instrument.

If the readout of a digital meter is steady for a given input signal, then the precision limit of the measurement could be as large as $\pm 1/2$ of the least digit in the output. Of course the precision limit could be significantly less than this value. When there is no flicker in the output of a digital instrument (at a steady-state condition), the precision errors are essentially damped by the digitizing process. Their magnitude is equal to or less than $\pm 1/2$ the least digit.

The significance of this potential precision error should be considered in the early design phase of an experiment. If the instrument resolution is such that a precision limit estimate of $\pm 1/2$ the least digit is unacceptably large, then a meter with a better resolution (more digits) should be considered for the experiment.

When data are obtained in the actual operation of the experiment, the precision limit used will be that observed in the meter readout. The precision limit will be $\pm 1/2$ of the flicker observed in the output. For a steady readout, a precision limit of $\pm 1/2$ the least digit should be used for the measurement.

In the absence of calibration, the instrument bias error should be obtained from the manufacturer's information. This accuracy specification will usually consider such factors as gain, linearity, and zero errors, and it should be taken as a bias error when no other information is available. This bias will be in addition to the digitizing bias and any transducer or system-sensor bias that may be present in the measurement.

The bias error in the measurement system can be reduced by calibration. However, the minimum bias limit will be either that associated with the calibration standard or $\pm 1/2$ of the least digit in the instrument output, whichever is larger.

5-3 DYNAMIC RESPONSE OF INSTRUMENT SYSTEMS

So far in our discussions, we have only considered errors in the output of instruments for static, or unchanging, inputs. Any time variations in the measured quantity have been treated as random variations that contribute to the precision error in the measurement. However, it is also necessary for us to consider the response of instruments to dynamic, or changing, inputs. An instrument may produce an output with both amplitude and phase ("time lag") errors when a dynamic input is encountered.

These dynamic response errors are similar to the variable but deterministic bias errors discussed in Section 5-1. In the following sections we will present the fundamentals needed to estimate these amplitude and phase errors.

5-3.1   General Instrument Response

The traditional way to investigate the dynamic response of an instrument is to consider the differential equation that describes the output. We will assume the instrument response can be modeled using a linear ordinary differential equation with constant coefficients [8]

$$a_n\frac{d^n y}{dt^n} + a_{n-1}\frac{d^{n-1}y}{dt^{n-1}} + \cdots + a_1\frac{dy}{dt} + a_0 y = bx(t) \qquad (5.40)$$

where y is the instrument output, x is the input, and n is the order of the instrument.

Instrument response to three different inputs will be discussed. The three are (1) a step change, (2) a ramp input, and (3) a sinusoidal input. These are illustrated in Figures 5.6, 5.7, and 5.8. Mathematically, these inputs are

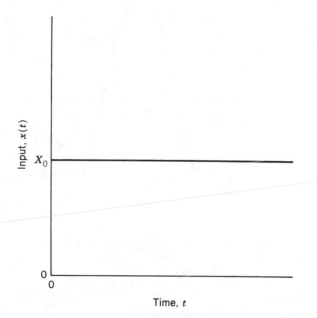

Figure 5.6   A step change in input to an instrument.

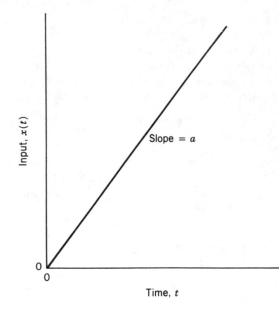

Figure 5.7   A ramp change in input to an instrument.

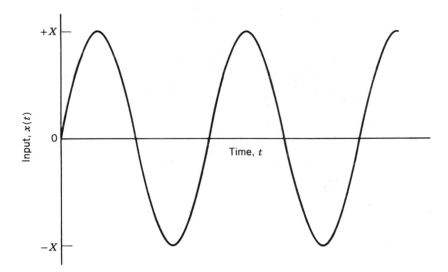

Figure 5.8   A sinusoidally varying input to an instrument.

described as follows:

1. step change $\quad\quad\quad x = 0 \quad\quad\quad\quad\quad t < 0$

$\quad\quad\quad\quad\quad\quad\quad\quad x = x_0 \quad\quad\quad\quad\quad t > 0 \quad\quad\quad\quad (5.41)$

2. ramp $\quad\quad\quad\quad\quad x = 0 \quad\quad\quad\quad\quad\quad t < 0$

$\quad\quad\quad\quad\quad\quad\quad\quad x = at \quad\quad\quad\quad\quad t \geq 0 \quad\quad\quad\quad (5.42)$

3. sinusoidal $\quad\quad\quad x = X\sin(\omega t) \quad\quad t > 0 \quad\quad\quad\quad (5.43)$

The response of zero-, first-, and second-order instruments to these inputs are considered next.

5-3.2 Response of Zero-Order Instruments

Since $n = 0$ for a zero-order instrument, Eq. (5.40) reduces to an algebraic equation

$$y = Kx(t) \quad\quad\quad\quad (5.44)$$

where K ($= b/a_0$) is called the static gain. Equation (5.44) shows that the output is always proportional to the input, so there is no error in the output due to the dynamic response. Of course, there will be static errors of the types we have previously discussed.

An example of a zero-order instrument is an electrical-resistance strain gauge. The input strain, ϵ, causes the gauge resistance to change by an incremental amount, ΔR, according to the relationship [9]

$$\Delta R = FR\epsilon \quad\quad\quad\quad (5.45)$$

where F is the gauge factor and R is the resistance of the gauge wire in the unstrained condition. Since the instrument itself, the gauge wire, is directly experiencing the input strain, there is no dynamic response error in the output.

5-3.3 Response of First-Order Instruments

The response equation for first-order instruments is usually written in the form

$$\tau\frac{dy}{dt} + y = Kx \quad\quad\quad\quad (5.46)$$

where τ ($= a_1/a_0$) is the time constant and K ($= b/a_0$) is the static gain. The definition of a first-order instrument is one that has a dynamic response behavior that can be expressed in the form of Eq. (5.46) [7].

A first-order instrument experiences a time delay between its output and a time varying input. An example is a thermometer or thermocouple that must

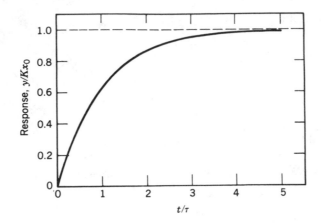

Figure 5.9 Response of a first-order instrument to a step change input versus nondimensional time.

undergo a heat transfer process for its reading to respond to a changing input temperature.

The response of a first-order instrument to a step change is found by solving Eq. (5.46) using Eq. (5.41) for x and the initial condition that $y = 0$ at $t = 0$. This solution can be expressed in the form

$$y/Kx_0 = 1 - e^{-t/\tau} \tag{5.47}$$

which is plotted in Figure 5.9. In one time constant, the response achieves 63.2% of its final value. One must wait for four time constants (4τ) before the response y will be within 2% of the final value.

The response to a ramp input is found by solving Eq. (5.46) using Eq. (5.42) for x and the initial condition that $y = 0$ at $t = 0$. This solution is

$$y = Ka\left[t - \tau(1 - e^{-t/\tau})\right] \tag{5.48}$$

which can also be expressed as

$$y - Kat = -Ka\tau(1 - e^{-t/\tau}) \tag{5.49}$$

Equation (5.48) is plotted in Figure 5.10. For no dynamic response error, we would obtain $y = Kat$ and the right-hand side (RHS) of Eq. (5.49) would be zero. The two terms on the RHS therefore represent the error in the response. The exponential term $(Ka\tau e^{-t/\tau})$ dies out with time and is called the transient error. The other term $(-Ka\tau)$ is constant and proportional to τ. The smaller the time constant is, the smaller this steady-state error will be. The effect of the steady-state error is that the output does not correspond to the input at the current time but to the input τ seconds before.

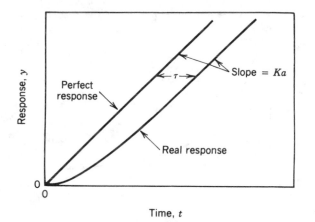

Figure 5.10   Response of a first-order instrument to a ramp input versus time.

The response of a first-order instrument to a sinusoidal input is found by solving Eq. (5.46) using Eq. (5.43) for x. This solution is

$$y = Ce^{-t/\tau} + \frac{KX}{\sqrt{1 + \omega^2\tau^2}} \sin(\omega t + \phi) \tag{5.50}$$

where

$$\phi = \tan^{-1}(-\omega\tau) \tag{5.51}$$

and C is the arbitrary constant of integration. The exponential term in Eq. (5.50) is the transient error that dies out in a few time constants. The second term on the RHS of Eq. (5.50) is the steady sinusoidal response of the instrument.

By comparing the steady response with the input [Eq. (5.43)], we see that the response has an amplitude error proportional to the amplitude coefficient $(1/\sqrt{1 + \omega^2\tau^2})$ and a phase error ϕ. These errors are shown in Figures 5.11 and 5.12. Each of these errors varies with the product of the time constant τ and the frequency of the input signal ω. As $\omega\tau$ increases, the amplitude coefficient decreases and the deviation from a perfect response becomes greater and greater as seen in Figure 5.11. A similar behavior is observed in Figure 5.12 for the phase error, which asymptotically approaches $-90°$ as $\omega\tau$ increases.

5-3.4   Response of Second-Order Instruments

With $n = 2$ in Eq. (5.40), the response of a second-order instrument becomes

$$a_2\frac{d^2y}{dt^2} + a_1\frac{dy}{dt} + a_0y = bx(t) \tag{5.52}$$

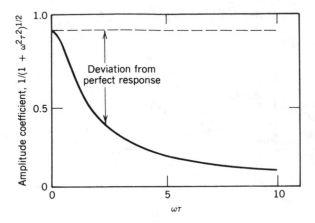

Figure 5.11   Amplitude response of a first-order instrument to a sinusoidal input of frequency ω.

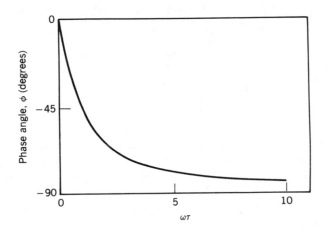

Figure 5.12   Phase error in the response of a first-order instrument to a sinusoidal input of frequency ω.

If this expression is divided through by a_2, it can be written in the form

$$\frac{d^2y}{dt^2} + 2\zeta\omega_n\frac{dy}{dt} + \omega_n^2 y = K\omega_n^2 x(t) \tag{5.53}$$

where $K \ (= b/a_0)$ is again the static gain, $\zeta \ (= a_1/2\sqrt{a_0 a_2})$ is the damping factor, and $\omega_n \ (= \sqrt{a_0/a_2})$ is the natural frequency. The definition of a second-order instrument is one that has a dynamic response behavior that can be expressed in the form of Eq. (5.53) [7]. Instruments that exhibit a spring-mass type of behavior are second order. Examples are galvanometers, accelerometers, diaphragm-type pressure transducers, and U-tube manometers [8].

The nature of the solutions to Eq. (5.53) is determined by the value of the damping constant ζ. For $\zeta < 1$, the system is said to be underdamped and the solution is oscillatory. For $\zeta = 1$, the system is critically damped, and for $\zeta > 1$ the system is said to be overdamped.

The second-order instrument response to a step change is found by solving Eq. (5.53) using Eq. (5.41) for x and the initial conditions that $y = y' = 0$ at $t = 0$. The solution depends on the value of ζ and is given by $\zeta > 1$:

$$y = Kx_0 \left\{ 1 - e^{-\zeta\omega_n t} \left[\cosh\left(\omega_n t \sqrt{\zeta^2 - 1}\right) \right. \right.$$

$$\left. \left. + \frac{\zeta}{\sqrt{\zeta^2 - 1}} \sinh\left(\omega_n t \sqrt{\zeta^2 - 1}\right) \right] \right\} \tag{5.54}$$

$\zeta = 1$:

$$y = Kx_0 \left[1 - e^{-\omega_n t}(1 + \omega_n t) \right] \tag{5.55}$$

$\zeta < 1$:

$$y = Kx_0 \left\{ 1 - e^{-\zeta\omega_n t} \left[\frac{1}{\sqrt{1 - \zeta^2}} \sin\left(\omega_n t \sqrt{1 - \zeta^2} + \phi\right) \right] \right\} \tag{5.56}$$

where

$$\phi = \sin^{-1}\left(\sqrt{1 - \zeta^2}\right) \tag{5.57}$$

This response is shown in Figure 5.13. Note that $1/\zeta\omega_n$ is now the "time constant." The larger $\zeta\omega_n$ is, the more quickly the response approaches the steady-state value. The form of the approach to the steady-state value is determined by ζ. For $\zeta < 1$, the response overshoots, then oscillates about the final value while being damped.

Most instruments are designed with damping factors about 0.7. The reason for this can be seen in Figure 5.13. If an overshoot of 5% is allowed, a damping factor $\zeta \cong 0.7$ will result in a response that is within 5% of the steady-state value in about half the time required by an instrument with $\zeta = 1$ [8]. Note that the steady-state solution for all values of $\zeta > 0$ gives Kx_0.

The response to a ramp input also contains a transient and steady-state portion. The steady-state solution is

$$y = Ka\left(t - \frac{2\zeta}{\omega_n}\right) \tag{5.58}$$

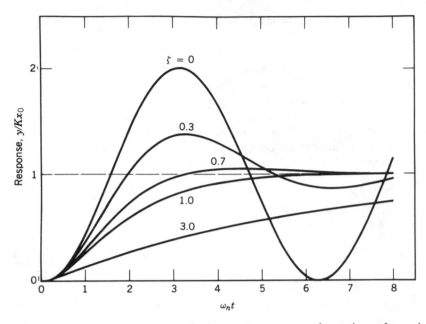

Figure 5.13 Response of a second-order instrument to a step change input for various damping factors.

The response lags behind the input by a time equal to $2\zeta/\omega_n$. High values of ω_n and/or low values of ζ reduce this lag in the steady-state response.

The response of a second-order instrument to a sinusoidal input is (at steady state) given by

$$y = \frac{KX}{\left[\left(1 - \omega^2/\omega_n^2\right)^2 + \left(2\zeta\omega/\omega_n\right)^2\right]^{1/2}} \sin(\omega t + \phi) \qquad (5.59)$$

where

$$\phi = \tan^{-1}\left(-\frac{2\zeta\omega/\omega_n}{1 - \omega^2/\omega_n^2}\right) \qquad (5.60)$$

As in the first-order system, the response contains both an amplitude error proportional to an amplitude coefficient and a phase error. These errors are shown in Figures 5.14 and 5.15.

From Figure 5.14 we see that for no damping ($\zeta = 0$), the amplitude of the response approaches infinity as the input signal frequency ω approaches the instrument natural frequency ω_n. In general this maximum amplitude, or

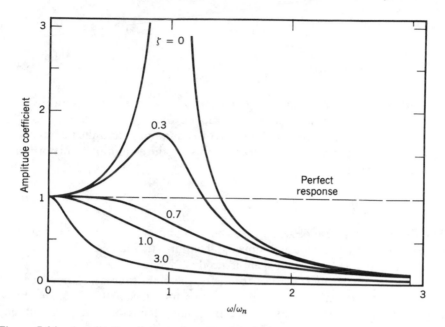

Figure 5.14 Amplitude response of a second-order instrument to a sinusoidal input of frequency ω.

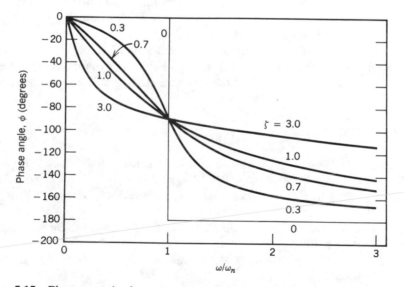

Figure 5.15 Phase error in the response of a second-order instrument to a sinusoidal input of frequency ω.

resonance, will occur at

$$\omega = \omega_n\sqrt{1 - 2\zeta^2} \qquad (5.61)$$

Note that for nonzero damping (which is always the case physically), the amplitude at this resonant frequency will be finite. Also note that for $\zeta \cong$ 0.6–0.7 and $\omega/\omega_n < 1$, the amplitude error is minimized and the phase error is about linear, which is desirable because this produces minimum distortion of the input signal [8].

5-3.5 Summary

The dynamic response of zero-, first-, and second-order instruments has been presented for step, ramp, and sinusoidal inputs. In the case of a zero-order instrument, it was found that there is no dynamic response error.

For first- and second-order instruments, there are time delays for step and ramp inputs and amplitude and phase errors for sinusoidal inputs. By choosing or designing instruments with appropriate values of time constant and natural frequency, the effects of these errors can be minimized.

In a complete measurement system, different instruments will usually be connected together in the form of a transducer, a signal conditioning device, and a readout. An example might be a thermocouple (a first-order instrument) connected to an analog voltmeter (a second-order instrument). In such cases the dynamic output of the first instrument can be determined and this value can be used as the input to the second instrument. The dynamic response of the second instrument to this input is then determined to obtain the dynamic response of the system. In most cases, the significant dynamic response error effect will occur with only one of the instruments—usually the transducer.

REFERENCES

1. Moffat, R. J., "Describing the Uncertainties in Experimental Results," *Experimental Thermal and Fluid Science*, Vol 1, Jan. 1988, pp 3–17.
2. Obert, E. F., *Internal Combustion Engines and Air Pollution*, Harper & Row, New York, 1973.
3. *Complete Temperature Measurement Handbook and Encyclopedia*, Omega Engineering Inc., 1987, p T37.
4. Incropera, F. P., and DeWitt, D. P., *Fundamentals of Heat Transfer*, John Wiley, New York, 1981.
5. Hottel, H. C., and Sarofim, A. F., *Radiative Transfer*, McGraw-Hill, New York, 1967.
6. Dally, J. W., Riley, W. F., and McConnell, K. G., *Instrumentation for Engineering Measurements*, John Wiley, New York, 1984.

7. Doebelin, E. O., *Measurement Systems Application and Design*, 3rd ed., McGraw-Hill, New York, 1983.

8. Schenck, H., *Theories of Engineering Experimentation*, 3rd ed., McGraw-Hill, New York, 1979.

9. Holman, J. P., *Experimental Methods for Engineers*, 4th ed., McGraw-Hill, New York, 1984.

CHAPTER 5

Problems

5.1 A thermocouple probe is located in the exhaust of a high temperature gas furnace. The digital thermometer connected to the probe indicates a temperature of 1925°C. The manufacturer specifies that the accuracy of the probe is ±1.5% of the temperature in °C and the accuracy of the meter is ±2°C. If the refractory wall temperature is 1100 ± 100°C and the convection coefficient between the gas and the probe is 200 ± 40 W/m²-K, what is the best estimate of the range that encloses the true gas temperature?

5.2 For the velocity profile example presented in Section 5-1.2, what Pitot tube diameter would be required to reduce the uncertainty at the center-line to no more than ±3%?

5.3 Compare the digitizing bias error for a signal of ±0.5 V or less for 8, 12, and 16 bit A/D converter systems for calibrated ranges of ±5 V and ±0.5 V.

5.4 Consider the calibration of an instrument system consisting of a trans-ducer connected to a signal conditioner with a built-in digital meter. The meter is calibrated for a range of ±100 mV with a least digit of 1 mV and the meter uses an 8-bit A/D converter. If the input signal to the meter is constant at 2 mV and if the calibration standard bias is less than the bias associated with the meter, estimate the precision and bias limits for this reading. (Be sure to consider the digitizing bias error.)

5.5 During a test, the reading on a digital meter varies randomly from 1.22 V to 1.26 V. The meter has a calibrated range of 0 to 10 V and uses a 12-bit A/D converter. The manufacturer specifies that the accuracy of the meter is ± (2% of the reading +1 times the least scale division). Estimate the bias and precision limits of the reading.

5.6 The response of a thermocouple to a step change in temperature can be modeled as

$$\frac{mC}{hA}\frac{d\theta}{dt} + \theta = \theta_\infty$$

where m, C, and A are the mass, specific heat, and surface area of the thermocouple, respectively, h is the heat transfer coefficient between the thermocouple and the fluid, θ is the difference between the thermocouple temperature, T, at time t and its initial temperature, T_0, and θ_∞ is the fluid temperature, T_∞, minus T_0. For a spherical thermocouple bead with $C = 385$ J/kg-K and density of 8900 kg/m^3 in a fluid with $h = 200$ W/m^2-K, what diameter would be required to have the response reach 98% of the final value in 20 sec?

5.7 For the thermocouple in Problem 5.6, what would the amplitude coefficient and phase error be for an input signal (fluid temperature) that varies with a frequency of 1 hertz? If the thermocouple diameter is increased, what effect will there be on its response to this signal? What must be done to the thermocouple system to get a transient response with a larger amplitude coefficient (less error)?

5.8 For an accelerometer with a damping factor of 0.7, what natural frequency should be specified if the signal to be measured has a frequency of 400 hertz and an error of no more than 2% is desired in the output amplitude?

6

DEBUGGING AND EXECUTION OF EXPERIMENTS

After proceeding through the planning and design phases of an experiment, choosing the instrumentation, and building up the experimental apparatus, typically a debugging and qualification phase is necessary. In this phase, we try to determine and fix the unanticipated problems that occur and reach the point at which we feel confident that the experiment is performing as anticipated. In the first section of this chapter we discuss making checks at the first-order level of replication to investigate the run-to-run scatter in timewise experiments. If our first-order precision limit estimate P_r accounts for all the significant sources of precision error, then the run-to-run scatter should fall within $\pm P_r$. We also discuss using the Nth-order uncertainty estimate (U) to compare results against accepted results or theoretical values. This can sometimes be done only for some limiting case or condition that covers a portion of the range of operation of the experiment, but nonetheless it serves as a valuable qualification check for the experimental effort.

The idea of using balance checks (or experimental agreement with conservation laws) in the debugging and execution phases of an experiment is discussed in Section 6-2, whereas the development and use of a "jitter program" to incorporate uncertainty analysis calculations in the data reduction program are discussed in Section 6-3.

Before entering the execution phase of the experiment, decisions must be made on which experimental set points are to be used and in what order. Ideas impacting these decisions are discussed in Section 6-4.

6-1 DEBUGGING AND QUALIFICATION CHECKS

6-1.1 Basic Ideas

The idea of different orders of replication level is very useful in the debugging and qualification stages of a timewise experiment. The way we have defined the first-order level of replication, its primary use is in checking the repeatability or scatter in the results r_i of a timewise experiment that is run more than once at a particular set point. The interval defined by $\bar{r} \pm P_r$ should enclose a result r_i 95 times out of 100. Here, P_r is the precision limit of the result determined from propagating the estimated precision limits of the measured variables using the uncertainty analysis expression. It is not the precision limit, $t\hat{S}_r$, determined from the results themselves.

The utility of such a comparison uses the following logic. If all the factors that influence the precision of the measured variables and the result have been properly accounted for in determining P_r, then the scatter in the results at a given set point should be approximated by $\pm P_r$. If the scatter in the results is greater than $\pm P_r$, this indicates that there are factors influencing the precision of the experiment that are not properly accounted for. This should be taken as an indication that additional debugging of the experimental apparatus and/or experimental procedures is called for.

The usefulness of the Nth-order replication level in the debugging/qualification of an experiment is in comparing current experimental results with theoretical results or previously reported, well-accepted results from other experimental programs. Since comparisons at the Nth-order level are viewed as comparisons of intervals within which "the truth" lies with 95% confidence, agreement should be expected within the overall uncertainty interval $(\pm U_r)$ of the result. The limits on this interval are

$$U_r = \left(P_r^2 + B_r^2 \right)^{1/2} \tag{6.1}$$

or

$$U_{\bar{r}} = \left(P_{\bar{r}}^2 + B_r^2 \right)^{1/2} \tag{6.2}$$

depending on whether one uses an individual result or an averaged result in the comparison. Agreement at the Nth-order level can be taken as an indication that all significant contributors to the uncertainty in the experimental result have been accounted for. Often this comparison, and hence the conclusion, can be made over only a portion of the operating domain of the experiment.

6-1.2 Example

As an example of an experimental program [1] that used checks at both the first- and Nth-order replication levels to advantage, consider the wind tunnel system shown in Figure 6.1. This system was designed to investigate the effect

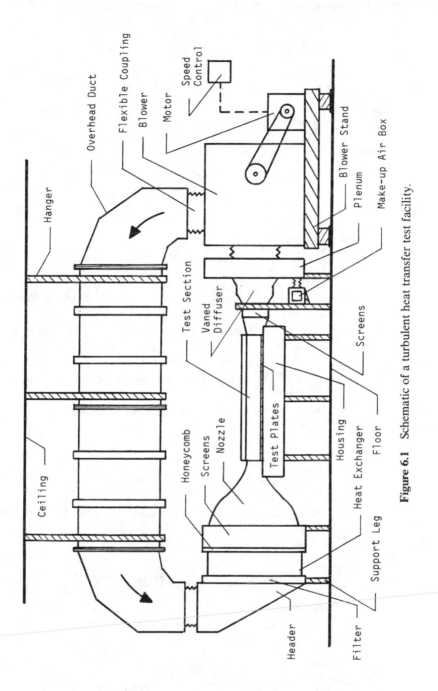

Figure 6.1 Schematic of a turbulent heat transfer test facility.

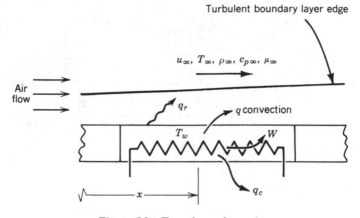

Figure 6.2 Test plate schematic.

of surface roughness on the heat transfer between a turbulent boundary layer air flow and a flat plate, which in this case was the bottom surface of the wind tunnel test section. The basic experimental approach was the steady-state technique described in Section 3-6. Twenty-four plate segments made up the test surface, each with its own electrical heater/power supply/instrumentation system. Each plate segment (Figure 6.2) was supplied the power, W, necessary to maintain its temperature at a constant prescribed value T_w. The convective heat transfer coefficient, h, was determined in nondimensional form as a Stanton number defined by

$$St = \frac{h}{\rho_\infty u_\infty c_{p\infty}} \tag{6.3}$$

where ρ_∞ is the air density, $c_{p\infty}$ the air constant pressure specific heat, and u_∞ the freestream air velocity.

Upon using the defining equation for h

$$q_{\text{convection}} = hA(T_w - T_\infty) \tag{6.4}$$

and an energy balance on a test plate segment that included conduction losses, q_c, and radiation losses, q_r, the data reduction equation became

$$St = \frac{W - q_r - q_c}{\rho_\infty u_\infty c_{p\infty} A(T_w - T_\infty)} \tag{6.5}$$

A test consisted of setting the desired value of freestream velocity, u_∞, and waiting until the computer-controlled data acquisition and control system adjusted the power to each plate so that all 24 plates were at a steady-state temperature T_w. Once such a steady state was reached, values of all variables were recorded and the data were reduced and plotted in log–log coordinates as St versus Re_x, where the Reynolds number is given by the second data reduction equation in the experiment

$$Re_x = \frac{\rho_\infty u_\infty x}{\mu_\infty} \tag{6.6}$$

and x is the distance from the leading edge of the test surface to the mid-point of a particular plate. The initial qualification testing was planned to use a smooth test surface and five values of u_∞ from 12 to 67 m/sec.

A detailed uncertainty analysis during the design phase using the techniques outlined in Chapter 4 concluded that precision errors in the determination of Stanton number were negligible compared to the bias errors, so that

$$P_{St} \approx 0 \tag{6.7}$$

and

$$B_{St} \approx U_{St} \approx 2\text{–}4\% \tag{6.8}$$

depending on the experimental set point.

This uncertainty analysis result indicated, then, that replications on different days at a particular set point (u_∞) should produce St data with negligible variation. That is, a check at the first-order replication level during the debugging phase should show negligible run-to-run scatter. Figures 6.3 and 6.4 show replicated data at $u_\infty = 12$ and 67 m/sec, respectively. These results support the conclusion of the detailed uncertainty analysis that $P_{St} \approx 0$. Figure 6.5, however, presents three replications at the $u_\infty = 57$ m/sec set point that indicate a problem. The St data for the 01/19/88 run are obviously different from the runs made on 01/21/88 and 01/26/88—the precision limit for St indicated by these three replications is definitely not zero. It therefore could not be concluded that the debugging phase was completed.

In this case, examination of the data logs for the three runs indicated a flaw in the experimental procedure and not in the instrumentation. In starting the experimental facility the morning of a run (which took from 4 to 8 hr to complete), the laboratory barometer was read and that value of barometric pressure entered into the computer program that was used for control and data acquisition. That same value of barometric pressure was used in data reduction. On January 19, 1988 a weather front moved through the area, causing severe thunderstorms, tornado warnings, and large excursions in barometric pressure over a time period of a few hours. Numerical experiments with the data reduction program and the 01/19/88 data showed that the discrepancy in the St data could be explained by the change in barometric pressure that occurred between the time the barometer was read and the time the rest of the variable values were read when steady state was achieved several hours later.

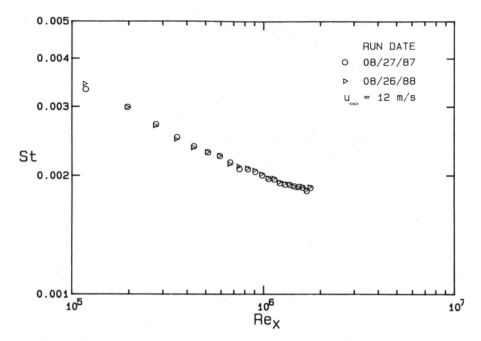

Figure 6.3 Debugging check at the first-order replication level. $u_\infty = 12$ m/sec.

In this experimental program, the check at the first-order replication level made during the debugging phase identified a flaw in the experimental procedure that could cause unanticipated variations in the experimental data. This was remedied by a change in the experimental procedure requiring the values of *all* variables to be determined and recorded within a period of several minutes once a steady state had been achieved.

Once a satisfactory conclusion had been reached at the first-order replication level, checks at the Nth-order replication level were made. Since neither accepted experimental data nor trustworthy theoretical results existed for the rough surface conditions to be investigated in the experimental program, the only "truth" available with which comparisons could be made at the Nth-order replication level was smooth surface St vs. Re_x data that had been reported previously and were widely accepted as valid. Such data had been reported by Reynolds, Kays, and Kline [2] in 1958 for Reynolds numbers up to 3,500,000. These data are shown in Figure 6.6 along with the power law correlation equation suggested by Kays and Crawford [3]

$$St = 0.0287(Re_x)^{-0.2}(Pr)^{-0.4} \qquad (6.9)$$

where Pr is the Prandtl number of the freestream air. As can be seen, an interval of $\pm 5\%$ about Eq. (6.9) encloses roughly 95% of the data. If these data are accepted as the standard, then we are essentially assuming with 95%

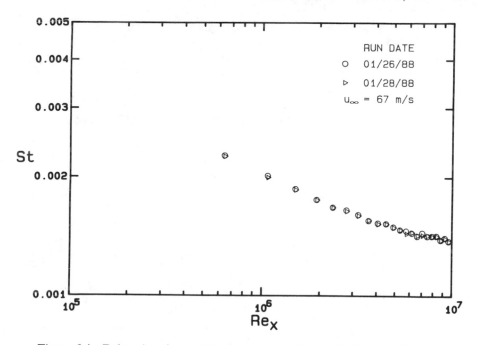

Figure 6.4   Debugging check at the first-order replication level. $u_\infty = 67$ m/sec.

confidence that the "true" Stanton numbers in the Reynolds number range from 100,000 to 3,500,000 lie in the $\pm 5\%$ band about Eq. (6.9).

Shown in Figure 6.7 is the qualification check at the Nth-order replication level. Data from the experiment are plotted along with uncertainty intervals ($\pm U_{St}$) for some representative data points. (The uncertainty interval on Re_x was less than $\pm 2\%$ and is therefore smaller than the width of the data symbols for the Re_x scale used in the figure.) In this case, the agreement is excellent in the sense that the new St data with the associated $\pm U_{St}$ intervals fall almost totally within the uncertainty band associated with the standard data. It was therefore concluded that a satisfactory qualification check for St at the Nth-order replication level had been made for $100,000 < Re_x < 3,500,000$. This then lent substantial credibility to the uncertainty estimates associated with the smooth wall St data for $Re_x > 3,500,000$ shown in Figure 6.7 and also to the subsequent data taken with rough surfaces in the same apparatus and using the same instrumentation.

6-2   USE OF BALANCE CHECKS

6-2.1   Basic Ideas

Balance checks are an application of the basic physical conservation laws (energy, mass, electrical current, etc.) to an experiment. The reasons for using balance checks are two-fold. First, in the debugging/qualification phase the

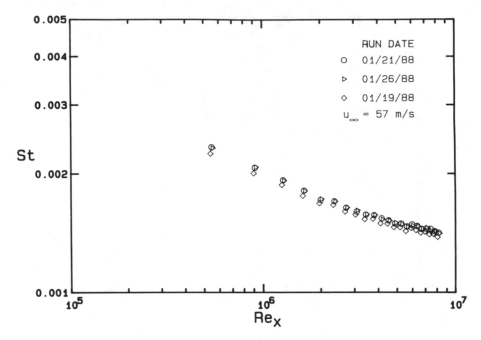

Figure 6.5 Debugging check at the first-order replication level. $u_\infty = 57$ m/sec. Problem indicated since $P_{St} \approx 0$ was expected.

application of a conservation law can help determine if errors exist that have not been taken into account. If a balance check is not satisfied within the estimated uncertainty interval (Nth order) for the measurements, then the debugging/qualification phase of the experiment cannot be assumed to be complete. Second, the use of balance checks that have been shown to be satisfied in the qualification phase can be useful in monitoring the execution phase of an experiment. If, for instance, a balance check suddenly fails to be satisfied, this can indicate an unrecognized change in instrument calibration or the process itself.

Balance checks may require more measurements than are needed by the basic experiment itself. Additional measurements must then be planned for in the design phase of the experiment.

6-2.2 Application to a Flow System

Consider the flow system in Figure 6.8. Three flowmeters are installed to measure the three mass flowrates $m1$, $m2$, and $m3$. If all the measurements were perfect, then conservation of mass would say that

$$m3 = m1 + m2 \tag{6.10}$$

However, uncertainties in the measurements (and possible leaks) make this

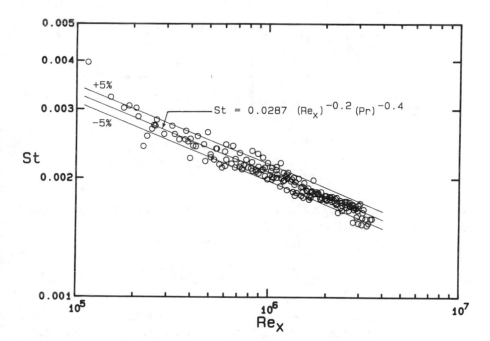

Figure 6.6 Classic data set of Reynolds, Kays, and Kline [2] widely accepted as a standard.

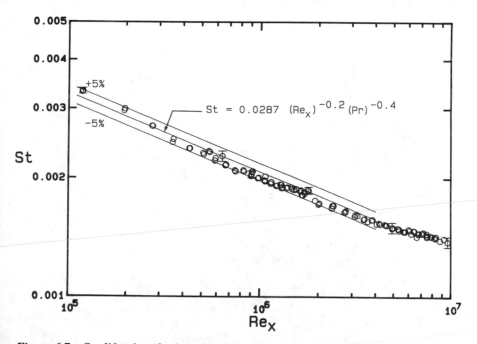

Figure 6.7 Qualification check at the Nth-order replication level. Successful comparison of new data with the standard set.

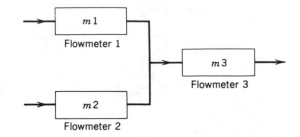

Figure 6.8   Flow system schematic for balance check example.

extremely unlikely. How closely must Eq. (6.10) be satisfied before we can conclude the balance check is satisfactory?

Take the quantity z as the mass balance parameter

$$z = m3 - m1 - m2 \tag{6.11}$$

Now the question becomes how close to zero must z be before we can say that the mass balance is satisfied? This must be answered if we are to use this mass balance to check our experiment in the debugging phase or monitor the experiment during the execution phase.

Considering z as an experimental result, the detailed uncertainty analysis expressions are

$$P_z^2 = \left(\frac{\partial z}{\partial m1} \right)^2 P_{m1}^2 + \left(\frac{\partial z}{\partial m2} \right)^2 P_{m2}^2 + \left(\frac{\partial z}{\partial m3} \right)^2 P_{m3}^2 \tag{6.12}$$

and

$$\begin{aligned}
B_z^2 &= \left(\frac{\partial z}{\partial m1} \right)^2 B_{m1}^2 + \left(\frac{\partial z}{\partial m2} \right)^2 B_{m2}^2 + \left(\frac{\partial z}{\partial m3} \right)^2 B_{m3}^2 \\[2mm]
&+ 2 \left(\frac{\partial z}{\partial m1} \right) \left(\frac{\partial z}{\partial m2} \right) B_{m1}' B_{m2}' \\[2mm]
&+ 2 \left(\frac{\partial z}{\partial m1} \right) \left(\frac{\partial z}{\partial m3} \right) B_{m1}' B_{m3}' \\[2mm]
&+ 2 \left(\frac{\partial z}{\partial m2} \right) \left(\frac{\partial z}{\partial m3} \right) B_{m2}' B_{m3}'
\end{aligned} \tag{6.13}$$

where, as usual, the $B_{mi}' B_{mj}'$ are the portions of the biases in mi and mj that are perfectly correlated. As

$$\frac{\partial z}{\partial m1} = \frac{\partial z}{\partial m2} = -1 \tag{6.14}$$

and

$$\frac{\partial z}{\partial m3} = +1 \tag{6.15}$$

Eqs. (6.12) and (6.13) become

$$P_z^2 = P_{m1}^2 + P_{m2}^2 + P_{m3}^2 \tag{6.16}$$

and

$$B_z^2 = B_{m1}^2 + B_{m2}^2 + B_{m3}^2 + 2B_{m1}'B_{m2}'$$
$$- 2B_{m1}'B_{m3}' - 2B_{m2}'B_{m3}' \tag{6.17}$$

and the overall 95% coverage uncertainty in z is given by

$$U_z^2 = P_z^2 + B_z^2 \tag{6.18}$$

For the balance check to be satisfied, the following must be true:

$$|z| \leq U_z \tag{6.19}$$

To investigate the behavior of U_z for various circumstances, assume for purposes of this example that all Ps and Bs are equal to 2 kg/hr in the following instances in which they are nonnegligible:

a. Precision dominated—all biases negligible:

$$U_z = \left(P_z^2 \right)^{1/2} = \left(P_{m1}^2 + P_{m2}^2 + P_{m3}^2 \right)^{1/2}$$
$$= [(4) + (4) + (4)]^{1/2} = 3.5 \text{ kg/hr}$$

Therefore, $|z| \leq 3.5$ kg/hr for the mass balance to be satisfied.

b. Precision errors negligible—all biases uncorrelated:

$$U_z = \left(B_z^2 \right)^{1/2} = \left(B_{m1}^2 + B_{m2}^2 + B_{m3}^2 \right)^{1/2}$$
$$= [(4) + (4) + (4)]^{1/2} = 3.5 \text{ kg/hr}$$

Therefore, $|z| \leq 3.5$ kg/hr for the mass balance to be satisfied.

c. Precision errors negligible—B_{m1} and B_{m2} wholly correlated, B_{m3} independent:

$$U_z = \left(B_z^2 \right)^{1/2} = \left(B_{m1}^2 + B_{m2}^2 + B_{m3}^2 + 2B_{m1}B_{m2} \right)^{1/2}$$
$$= [(4) + (4) + (4) + (2)(2)(2)]^{1/2} = 4.5 \text{ kg/hr}$$

Therefore, $|z| \leq 4.5$ kg/hr for the mass balance to be satisfied. This case could occur if all bias in $m1$ and $m2$ was calibration bias and both were calibrated against the same standard, while meter 3 was calibrated against a different standard.

d. Precision errors negligible—B_{m1}, B_{m2}, and B_{m3} wholly correlated:

$$U_z = \left(B_z^2 \right)^{1/2}$$

$$= \left(B_{m1}^2 + B_{m2}^2 + B_{m3}^2 + 2B_{m1}B_{m2} - 2B_{m1}B_{m3} - 2B_{m2}B_{m3} \right)^{1/2}$$

$$= [(4) + (4) + (4) + (2)(2)(2) - (2)(2)(2) - (2)(2)(2)]^{1/2}$$

$$= 2.0 \text{ kg/hr}$$

Therefore, $|z| \leq 2.0$ kg/hr for the mass balance to be satisfied. This case could occur if all biases were calibration biases and all three meters were calibrated against the same standard.

e. Precision and bias important—B_{m1} and B_{m2} wholly correlated, B_{m3} independent:

$$U_z = \left(P_z^2 + B_z^2 \right)^{1/2}$$

$$= \left(P_{m1}^2 + P_{m2}^2 + P_{m3}^2 + B_{m1}^2 + B_{m2}^2 + B_{m3}^2 + 2B_{m1}B_{m2} \right)^{1/2}$$

$$= [(4) + (4) + (4) + (4) + (4) + (4) + (2)(2)(2)]^{1/2}$$

$$= 5.7 \text{ kg/hr}$$

Therefore, $|z| \leq 5.7$ kg/hr for the mass balance to be satisfied.

In summary for this example, we have seen the general approach to formulating a balance check using uncertainty analysis principles. We also note the possibility of correlated biases playing a major role in determining the degree of closure required for the balance to be satisfied.

6-3 USE OF JITTER PROGRAM

A jitter program can be an extremely useful tool and should be implemented in the debugging/qualification and execution phases of almost every experimental program. In a jitter program, the data reduction computer program is treated as a subroutine that is successively iterated so that all of the partial derivatives necessary in uncertainty analysis are calculated using finite difference approximations. Moffat [4, 5, 6] introduced this idea for propagating general uncertainty intervals into a result. Here, we extend the idea to include the separate propagation of bias limits and precision limits into a result and also to include the effects of correlated bias limits.

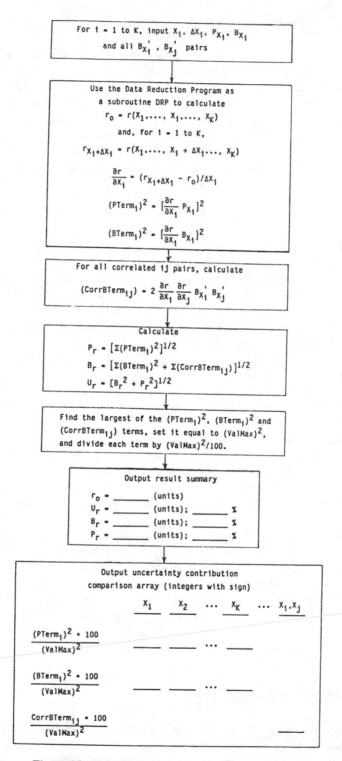

Figure 6.9 Schematic of jitter program procedure.

The biggest advantage in using a jitter program is that whenever a change in the data reduction program is made, the uncertainty analysis program is automatically updated since the data reduction program itself is used to generate the partial derivative values. In complex experiments with extended debugging/qualification phases, this is of great importance. The ease with which the jitter program can be implemented, however, practically dictates its use even in experiments with very simple data reduction programs.

The jitter program procedure is outlined in Figure 6.9 for a case in which the experimental result r is a function of K variables X_i. The X_is include not only the variables that are measured, but also those (such as emissivity, for example) whose values are usually found from a reference source. These latter variables should be included in the data reduction program (DRP) as named variables, not as a single numerical value, as they must be perturbed in the jitter program so that a partial derivative can be calculated.

As seen in the figure the inputs required are the data itself (the values of the X_is), the estimates of the precision limit and bias limit for each variable, the estimates of the correlated bias pairs (if any), and specification of the amount (ΔX_i) each variable is to be perturbed in the finite difference approximation of the partial derivatives. The DRP is used $K + 1$ times to compute the result r_0 and a perturbed result $r_{X_i + \Delta X_i}$ for each variable, and the finite difference approximations to the partial derivatives are computed together with the terms for the P_r and B_r propagation equations. At this point P_r, B_r, and U_r are calculated and each of the terms contributing to P_r^2 and B_r^2 is normalized so that the largest term has a value of 100.

A suggested output format is shown in the figure. The result r_0 is (obviously) output, and U_r, B_r, and P_r are output both in dimensional form and as percentage of r_0. The percentage output form is especially useful for allowing a quick comprehension of the importance of the uncertainty components when the result and/or U_r, B_r, and P_r are either very large or very small numbers relative to unity.

The second part of the output has been found to be very useful by the authors. The array of normalized values of the individual contributors to B_r^2 and P_r^2 contains numbers ranging from 0 to 100, with the correlated bias terms having the possibility of being negative as well as positive. One can quickly glance over this array and discern which contributors are important and which are negligible. This is particularly important to track in experiments that cover a broad range of conditions. In such cases, a contributor might well be negligible in one regime but be dominant in another.

6-4   EXPERIMENT EXECUTION

In the execution phase of an experimental program, the results are obtained for various "set" values of the measured variables. These set points should be chosen carefully so that the experiment will provide uniform information about the results over the full range of expected values. In the first part of this section we will consider the concept of rectification and its use in test point spacing.

Once the test points are chosen, the order in which they are to be run must be decided. As we will see, a random order of test point settings rather than a sequential order is preferred.

6-4.1   Choice of Test Points: Rectification

The spacing of the test data points is achieved by choosing the value of the controlled variables that will be "set." The values of the dependent variables are determined by measurement at each of the settings of the controlled variables. These test points should be chosen with careful thought before the experiment is begun. If the relationship between the dependent variable and the controlled variable is nonlinear, an equal spacing of the controlled variable settings over the range of interest will lead to an unequal spacing in the dependent variable. This may lead to more data points than needed in one range of the variables and two few points in another range. The idea of rectification is key in the choice of coordinate systems that should be considered when determining test point spacing.

Assume that we are dealing with an experiment that produces data of high precision—that is, groups of data pairs (x, y) that give a smooth curve with relatively little scatter when plotted as y versus x. With high precision data, we try to use the coordinates that will rectify the data, or cause the points to fall on a straight line. Note that here x and y might very well be nondimensional parameters consisting of several different measured variables.

The data, when plotted in a straight line, can be represented by the equation

$$Y = aX + b \qquad (6.20)$$

a. For data pairs (x, y) whose functional relationship is

$$y = ax + b \qquad (6.21)$$

we see that $Y = y$ and $X = x$ and we obtain a straight line by plotting y vs. x on linear–linear paper.

b. For data pairs (x, y) whose functional relationship is

$$y = cx^d \qquad (6.22)$$

we can take the log of (6.22) and obtain

$$\log y = \log c + d \log x \qquad (6.23)$$

so that, comparing with Eq. (6.20), we have $Y = \log y$, $X = \log x$, $b = \log c$ and $a = d$. Therefore, if we plot $\log y$ versus $\log x$ on linear–linear paper, we will obtain a straight line. The same result can be obtained by plotting y versus x on log–log paper.

  c. For data pairs (x, y) whose functional relationship is

$$y = ce^{dx} \tag{6.24}$$

we can take the natural logarithm (ln) of Eq. (6.24) and obtain

$$\ln y = \ln c + dx \tag{6.25}$$

so that, comparing with Eq. (6.20), we have $Y = \ln y$, $X = x$, $a = d$, and $b = \ln c$. Therefore, if we plot $\ln y$ versus x on linear–linear paper, we will obtain a straight line. Plotting y versus x on semilog paper also produces a straight line.

  These three cases are the most common that occur in engineering. How do we use this information? If we know the general functional relationship between the variables before running an experiment, we should plan to present the data in the coordinates that rectify the data.

  In many cases, the general form of the functional relationship between the variables is known, and this can be used to determine test point spacing. The following two cases illustrate this point.

Case 1. In a certain experiment X is the controlled variable and Y is the result. Previous results in similar physical cases indicate a probable functional relationship of the form

$$Y = aX^b \tag{6.26}$$

We intend to curvefit the data to obtain the "best" values of the constants a and b. What spacing of X values should be used? Note that we must choose the data point spacing before we know the values of a and b.

Solution. If we take the logarithm of both sides of (6.26) we obtain

$$\log Y = \log a + b \log X \tag{6.27}$$

From Eq. (6.27) we find that if we consider $(\log Y)$ and $(\log X)$ as our variables, we have a linear relationship irrespective of the exponent of X. We can therefore use equal increments of $(\log X)$ as our set points to obtain equally spaced values of $(\log Y)$.

Case 2. Same statement as in Case 1, except the probable relationship is of the form

$$Y = ae^{-bX} \tag{6.28}$$

Solution. Again taking logarithms we obtain

$$\ln Y = \ln a - bX \tag{6.29}$$

and we see that using equally spaced values of X and considering ($\ln Y$) as the dependent variable will lead to evenly spaced data points.

Another factor that should be taken into account when determining test point spacing is the expected uncertainty in the result over the range of interest. If uncertainty analysis has indicated a higher precision limit contribution to the uncertainty in one portion of the range than another, one might want to take more points in the region in which the uncertainty is highest, remembering that

$$S_{\bar{y}} = \hat{S}_y / \sqrt{N} \tag{6.30}$$

Therefore, as a rule of thumb, four times the number of points gives a two times decrease in the precision limit contribution to the uncertainty.

6-4.2 Example of Use of Rectification

An experimental water flow facility to determine pipe friction factors for various Reynolds numbers is shown in Figure 6.10. The pump speed is set so that the desired Reynolds number

$$\text{Re} = \frac{4\rho Q}{\pi \mu D} \tag{6.31}$$

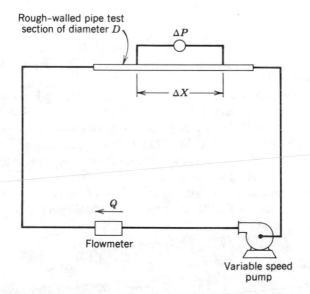

Figure 6.10 Schematic of flow loop for friction factor determination.

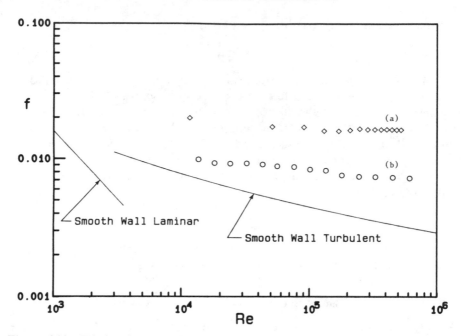

Figure 6.11   Friction factor vs. Reynolds number data for two rough wall pipes: (a) points equispaced in Re, (b) points equispaced in log(Re).

is achieved, and then the friction factor

$$f = \frac{\pi^2 D^5 \Delta P}{32 \rho Q^2 \Delta X}$$

(6.32)

is determined from measurements of ΔP and Q. It is known from previous experiments that for a given pipe roughness, the friction factor is a function of Reynolds number.

Such data are traditionally presented in a $\log(f)$ vs. $\log(\mathrm{Re})$ format—recall the famous "Moody diagram" from basic fluid mechanics. Shown in Figure 6.11 are the results of testing two test sections with rough walls for 14 Reynolds numbers from 13,000 to 611,000. The upper data set was run with set points equispaced in Re, and the lower set was run with set points equispaced in log(Re). The advantage of choosing set points with equal spacing in the coordinates of data presentation is obvious.

The determination of the Re set points for equal spacing in log(Re) is as follows:

$$\left. \begin{array}{l} \mathrm{Re} = 611{,}000 \rightarrow \log \mathrm{Re} = 5.786 \\ \mathrm{Re} = 13{,}000 \rightarrow \log \mathrm{Re} = 4.114 \end{array} \right\} \Delta = 1.672$$

divide Δ by $N - 1 = 13 \rightarrow 0.1286$

therefore $(\log \mathrm{Re})_i = (\log \mathrm{Re})_{i-1} + 0.1286$ (for $i = 2{-}14$)

i	log Re	Re
1	4.1140	13,002
2	4.2426	17,482
3	4.3712	23,507
4	4.4998	31,608
5	4.6284	42,501
6	4.7570	57,147
7	4.8856	76,842
8	5.0142	103,323
9	5.1428	138,931
10	5.2714	186,810
11	5.4000	251,189
12	5.5286	337,754
13	5.6572	454,151
14	5.7858	610,661

In this example we have illustrated the use of rectification to determine test point spacing for the case in which both the dependent variable (f) and the independent variable (Re) are results determined from measured quantities (Q, D, ΔP, and ΔX) and property values (μ and ρ). These two results are appropriate dimensionless groups for presenting this type of data. By arranging the measured values and the property data into dimensionless groups, we have reduced the number of variables that must be varied in the experiment.

To set each of the Reynolds numbers specified as test points, we can vary either the pipe diameter (D), the flow rate (Q), or the fluid itself (ρ and μ). In an actual experiment we would probably keep the fluid and the pipe diameter fixed and vary the flow rate. These results would then be general and would apply for any other flow situations that had the same Reynolds number and nondimensional roughness.

The appropriate dimensionless groups should be considered in any test plan. Many references are available on dimensional analysis. In general its use will make the experiment easier to conduct and will make the results more universally applicable.

6-4.3   Test Sequence

There are many arguments for choosing a random order of test point settings rather than a sequential order (from lowest to highest value, or vice versa). Among these are [7]

a. Natural effects—uncontrolled (or "extraneous") variables that might have a small but nevertheless measurable effect on the results and that may follow a general trend during the test. Examples are relative humidity and barometric pressure.

b. Human activities—the operator/data recorder may become more proficient during the test or may become bored and careless.

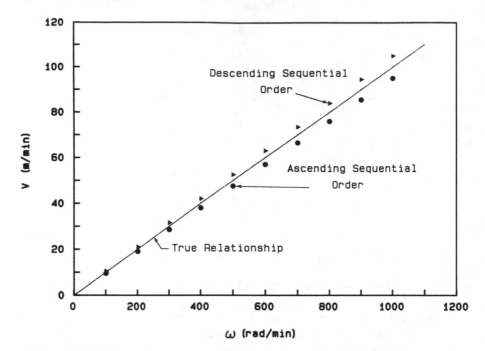

Figure 6.12   Data taken in sequential order and including 5% hysteresis effect.

c. Hysteresis effects—instruments may read high if the previous reading was higher and low if the previous reading was lower.

When possible, random sequences should be used when conducting an experiment. As an example, suppose that a person is measuring the velocity V of a point P on the edge of a spinning disk of radius $r = 10$ cm as a function of ω. The person is unaware of the relationship

$$V = r\omega \qquad (6.33)$$

Suppose further that the only errors are bias errors due to hysteresis in the velocity measurement system, which reads 5% low if the set point is approached from below and 5% high if the set point is approached from above.

Shown in Figure 6.12 are one data set taken in a sequentially increasing manner and one data set taken in a sequentially decreasing manner compared with the true relationship. It is obvious that the "best" line through either data set does not give a good picture of the true relationship.

Shown in Figure 6.13 is a data set with the points taken in a random order, with some points approached from above and some from below. Although the points in this data set scatter more than the sets in Figure 6.12, they scatter about the true relationship and thus give a better idea of the true system behavior.

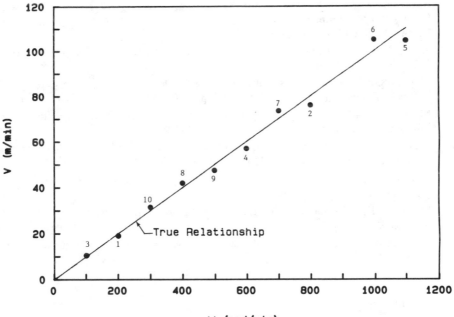

Figure 6.13   Data taken in random order and including 5% hysteresis effect.

In a general experiment in which the result r is a function of several variables

$$r = f(x, y, z) \qquad (6.34)$$

the test plan will consist of holding all of the independent variables but one fixed and varying that one over its full range. The process is then repeated for the other independent variables. This approach is referred to as a classical test plan and is the common method used for most engineering experiments [7].

There are other test plan techniques that can minimize the number of conditions that must be run while still providing statistically valid information. These are factorial plans in which all of the variables in the experiment are varied simultaneously. Discussion of these approaches is beyond the scope of this book, and the reader is referred to Hicks [8] and Montgomery [9] for more information on this type of test plan.

REFERENCES

1. Coleman, H. W., Hosni, M. H., Taylor, R. P., and Brown, G. B., "Smooth Wall Qualification of a Turbulent Heat Transfer Test Facility," Report TFD-88-2, Mechanical and Nuclear Engineering Department, Mississippi State University, 1988.

2. Reynolds, W. C., Kays, W. M., and Kline, S. J., "Heat Transfer in the Turbulent Incompressible Boundary Layer, Parts I, II, and III," NASA MEMO 12-1-58W, 12-2-58W, and 12-3-58W, 1958.

3. Kays, W. M., and Crawford, M. E., *Convective Heat and Mass Transfer*, McGraw-Hill, New York, 1980.

4. Moffat, R. J., "Contributions to the Theory of Single-Sample Uncertainty Analysis," *J. Fluids Engineering*, Vol 104, June 1982, pp 250–260.

5. Moffat, R. J., "Using Uncertainty Analysis in the Planning of an Experiment," *J. Fluids Engineering*, Vol 107, June 1985, pp 173–178.

6. Moffat, R. J., "Describing the Uncertainties in Experimental Results," *Experimental Thermal and Fluid Science*, Vol 1, Jan. 1988, pp 3–17.

7. Schenck, H., *Theories of Engineering Experimentation*, 3rd ed., McGraw-Hill, New York, 1979.

8. Hicks, C. R., *Fundamental Concepts in the Design of Experiments*, 3rd ed., Holt, Rinehart & Winston, New York, 1982.

9. Montgomery, D. C., *Design and Analysis of Experiments*, 2nd ed., John Wiley, New York, 1984.

CHAPTER 6

Problems

6.1 A heat exchanger, which uses water to cool the oil used in a transmission, is being tested to determine its effectiveness. An energy balance on the heat exchanger (assuming negligible energy loss to the environment) yields

$$\dot{m}_0 c_0 (T_{0,i} - T_{0,0}) = \dot{m}_w c_w (T_{w,0} - T_{w,i})$$

where

$\dot{m}_0$ = mass flow rate of oil

c_0 = specific heat of oil

$T_{0,i}$ = oil temperature at oil inlet

$T_{0,0}$ = oil temperature at oil outlet

$\dot{m}_w$ = mass flow rate of water

c_w = specific heat of water

$T_{w,0}$ = water temperature at water outlet

$T_{w,i}$ = water temperature at water inlet

Assume that the entire experiment is bias dominated—that is, the precision errors are negligible relative to bias errors. The mass flow rate measurements have (uncorrelated) bias limits of 1% of reading, and the values of specific heat have fossilized bias limits estimated as 1.5%. All temperature measurements have an elemental bias limit contribution of 1.0°C due to the nonuniformity of fluid temperatures across the flow inlets and outlets. All temperature measurements also have calibration bias limits of 1.5°C.

For nominal conditions such that

$$T_{0,i} - T_{0,0} = 40°C$$

and

$$T_{w,0} - T_{w,i} = 30°C,$$

determine the bounds within which an energy balance should be expected to close for

(a) all biases uncorrelated

(b) all calibration biases in the temperature sensors correlated.

6.2 In a particular experiment, the dc electrical power supplied to a circuit component must be measured. A dc watt transducer that the manufacturer says can measure the power to within 0.4% of reading is to be used. As a check during debugging, the voltage drop (E), current (I), and resistance (R) of the component are to be measured with a multimeter that has uncertainty specifications of 0.1% for E, 0.3% for R, and 0.75% for I. Nominal values are 1.7 A, 12 Ω, and 20.4 V. Denote W_1 as the power indicated by the meter output, $W_2 = EI$, and $W_3 = I^2R$. Assume all precision limits negligible.

To consider the balance check satisfied, how closely should W_1 and W_2 agree? W_1 and W_3? W_2 and W_3? W_1, W_2, and W_3?

6.3 For a centrifugal pump, the power input required, p, for a given pump speed, N, is governed by

$$p = CN^3$$

A range of speeds from 500 to 2000 rpm is being investigated, and 6 points are to be run (including the end points). A linear representation with equal spacing of points is desired. How should the data be presented and what N values should be run?

6.4 The attenuation of a radiation beam as it passes through a medium can be expressed as

$$I = I_0 e^{-\beta x}$$

where I_0 is the intensity of the beam entering the medium and I is the intensity at a depth x. β is a parameter that depends on the type of radiation and the medium. For a specific medium, in which β will probably be a constant, it is planned to measure the intensity of gamma rays at different depths where x will vary from 0.1 to 2 m. Seven points are to be run (including the end points) and a linear representation with equally spaced points is desired. How should the data be presented and what x locations should be used?

6.5 The heat transfer coefficient for freely falling liquid droplets can be expressed as

$$\mathrm{Nu} = a + b\,\mathrm{Re}^c$$

where a, b, and c are constants and Nu and Re are the Nusselt and Reynolds numbers, respectively. If the Reynolds number ranges up to 350, what Re values should be considered for test measurements if 10 data points are to be obtained? How should these data be represented graphically?

7

DATA ANALYSIS AND REPORTING OF RESULTS

Once data are acquired during an experiment, we must decide how the data can be most effectively presented. Much of the effort that has gone into planning, designing, and running the experiment will be wasted if the results are presented only in tabular form. The relationships between variables and the scatter in the data are usually very difficult to see from tabular data listings. Usually, graphic and mathematical representations of the data are used in addition to tabular presentation.

As a minimum, a report presenting experimental data should contain the following tabular information for each result: the value (single reading or average), the bias limit, the precision limit, and the uncertainty with the confidence level indicated. The data also should usually be presented in graphic form with the uncertainty bands around the results clearly indicated. We also often wish to obtain and report a mathematical expression (or "correlation equation") to represent the data we have obtained. An equation is a much more compact representation of the data than tabular listings or even graphic representations, and it allows us to interpolate between the discrete data points.

In the following sections, we consider how to obtain a mathematical expression to represent experimental results and what sort of uncertainty should be associated with such an expression. We then illustrate, using an example, how experimental results (and the uncertainties) might be presented in tabular and graphic form and also represented using a regression equation. The final section discusses ideas associated with multiple linear regression.

7-1 REGRESSION (CURVEFITTING)

Suppose that we wish to represent a set of experimental results with a mathematical expression. We should ask the question "What is the best equation?," or, put another way, "What is the best line through the data?" Often the "best" line through the data points is defined as that which makes the sum of the squares of the deviations of the data points from the line a minimum. This leads to the well-known *least-squares* method of curvefitting.

Consider a set of data points presented on an $X–Y$ plane. X and Y may be variables in the experiment, some function of the variables ($Y = \ln y$, for instance), or nondimensional groups of variables (such as Nu and GrPr). Assume that all the uncertainty is concentrated in Y and that the uncertainty in the Y variable does not vary with the value of Y. Also assume that the "true" relationship between X and Y is a straight line. This is not too serious a restriction when one considers the ideas concerning the rectification of data that were discussed in Chapter 6.

We can now assume the general linear equation

$$Y_0 = aX + b \tag{7.1}$$

where Y_0 is taken as the "optimum" Y value to represent the data at a given X. We wish to find the values of the slope, a, and intercept, b, that minimize

$$\eta = \sum_{i=1}^{N} (Y_i - Y_0)^2 \tag{7.2}$$

where the Y_i are the N data values. Using Eq. (7.1), the expression to minimize becomes

$$\eta = \sum_{i=1}^{N} (Y_i - aX_i - b)^2 \tag{7.3}$$

For η to be minimum, we must find a and b such that

$$\frac{\partial \eta}{\partial a} = 0 \tag{7.4}$$

and

$$\frac{\partial \eta}{\partial b} = 0 \tag{7.5}$$

Taking the indicated derivatives and solving for a and b leads to

$$a = \frac{N\Sigma X_iY_i - \Sigma X_i \Sigma Y_i}{N\Sigma(X_i^2) - (\Sigma X_i)^2} \tag{7.6}$$

and

$$b = \frac{\Sigma(X_i^2)\Sigma Y_i - \Sigma X_i \Sigma(X_iY_i)}{N\Sigma(X_i^2) - (\Sigma X_i)^2} \tag{7.7}$$

where the summations run from $i = 1$ to $i = N$.

The equation that results from this procedure

$$Y = aX + b \tag{7.8}$$

is called a *regression equation*, and the procedure itself is sometimes referred to as "making a regression of Y on X." One point that should be emphasized concerns the intercept, b. In many cases, we know what the intercept should be if the experiment were perfect. Often this "perfect" intercept is $Y = 0$ at $X = 0$. This "perfect" value of the intercept should never be forced on the data. As Schenck [1] states, "The intercept of straight-line data is always inherent in the data and should be allowed to express itself."

7-2   REGRESSION UNCERTAINTY

Consider that a linear least-squares curvefit [Eq. (7.8)] has been made for a set of data. We can then form a statistic called the standard error of estimate (SEE) [2]

$$SEE = \left\{ \frac{\sum_{i=1}^{N} [Y_i - (aX_i + b)]^2}{N - 2} \right\}^{1/2} \tag{7.9}$$

that has the general form of a precision index and that is a measure of the scatter of the data points about the linear curvefit. The 2 subtracted from N in the denominator arises because two degrees of freedom are lost from the set of N data pairs (X_i, Y_i) when the curvefit constants a and b are determined by least squares.

Schenck [1] states that a $\pm 2(SEE)$ band around the curvefit will contain approximately 95% of the data points. He proposes to call this band a "confidence interval" on the curvefit. However, this can be considered reasonable only for a case in which the bias contribution to the uncertainty is negligible.

A more general statement about the uncertainty that should be associated with a value Y determined from a curvefit is that the uncertainty should reflect the uncertainty associated with the data points used in the regression. The uncertainty band around each data point (result) represents the range in which the true value lies with 95% confidence. Therefore, a comparison of the curvefit and the data point uncertainty limits will allow us to determine the appropriate uncertainty to be associated with a value calculated from the curvefit.

7-3   EXAMPLE OF PRESENTATION OF EXPERIMENTAL RESULTS

In this example we discuss and illustrate the tabular, graphic, and mathematical presentation of results from an experiment dealing with free convection heat transfer from horizontal cylinders of finite length. Two different L/D (length/diameter) ratios were considered—a 1/8-in.-diameter cylinder 6 in. long ($L/D = 48$) and a 1-in.-diameter cylinder 6 in. long ($L/D = 6$). The experiment was conducted using the steady-state technique described in Section 3-6.2. The cylinders were electrically heated and the difference between the cylinder surface temperature and the air temperature was measured directly with a thermocouple circuit. For this case of no forced air flow, radiation effects were important and were considered in the data reduction. The precision errors for this experiment were estimated to be negligible with respect to the bias errors, and this was found to be the case when the experiment was performed and several set points were replicated.

7-3.1   Presentation of Results

The experimental data for the two cylinders are presented in Tables 7.1 and 7.2 in both dimensional and nondimensional form. Based on previously reported results in the literature, we anticipated the appropriate dimensionless groups for this experiment to be the Nusselt number, the product of the Grashof and Prandtl numbers, and the cylinder length to diameter ratio, L/D. The Nusselt number, Nu, is defined as

$$\mathrm{Nu} = \frac{hD}{k} \qquad (7.10)$$

where h is the heat transfer coefficient, D is the cylinder diameter, and k is the fluid thermal conductivity. The Grashof–Prandtl product GrPr is defined as

$$\mathrm{GrPr} = \frac{c_p \rho^2 g \beta D^3 \Delta T}{\mu k} \qquad (7.11)$$

where c_p, ρ, μ, and β are the fluid properties of specific heat, density,

TABLE 7.1   Experimental Results for Free Convection Heat Transfer from a Horizontal Cylinder ($L/D = 48$)[a]

Surface-to-Air Temperature Difference ΔT (F)	Convective Heat Transfer Coefficient h (Btu/hr-ft²-F)	Uncertainty U_h (Btu/hr-ft²-F)	Grashof–Prandtl Number GrPr	Nusselt Number Nu	Uncertainty U_{Nu}
7.9	2.13	0.28	12.1	1.44	0.19
8.1	2.06	0.26	12.4	1.39	0.18
11.4	2.26	0.21	17.5	1.53	0.14
11.4	2.24	0.21	17.5	1.51	0.14
16.6	2.43	0.16	25.6	1.64	0.11
21.2	2.53	0.14	32.6	1.71	0.10
21.4	2.58	0.14	33.0	1.75	0.10
27.1	2.67	0.12	41.7	1.81	0.09
32.4	2.77	0.11	49.8	1.88	0.09
32.8	2.76	0.11	50.5	1.87	0.08
37.6	2.88	0.11	57.9	1.95	0.08
46.0	2.94	0.10	70.7	1.99	0.08

[a] Uncertainty values are for $U_h \approx B_h$ and are at a 95% confidence level.

viscosity, and volumetric thermal expansion coefficient respectively, g is the acceleration of gravity, and ΔT is the difference between the cylinder surface temperature and the fluid temperature.

Although it is important to present the test results in tabular form as in Tables 7.1 and 7.2, a graphic representation is much more useful for observing the full implications of the results. The h values are plotted versus ΔT in Figure 7.1 for the cylinder with $L/D = 48$. The 95% confidence uncertainty band is shown for each data point.

Also shown in the figure is a curve representing the classic results for cylinders with large L/D ($L/D \to \infty$). The regression equation for this correlation curve is [3]

$$Nu = C(GrPr)^n \qquad (7.12)$$

where

GrPr	C	n
10^{-2}–10^2	1.02	0.148
10^2–10^4	0.850	0.188
10^4–10^7	0.480	0.250

After reviewing the data used to obtain this correlation equation, we con-

TABLE 7.2   Experimental Results for Free Convection Heat Transfer from a Horizontal Cylinder ($L / D = 6$)[a]

Surface-to-Air Temperature Difference ΔT (F)	Convective Heat Transfer Coefficient h (Btu/hr-ft²-F)	Uncertainty U_h (Btu/hr-ft²-F)	Grashof–Prandtl Number GrPr	Nusselt Number Nu	Uncertainty U_{Nu}
3.4	1.06	0.31	2640	5.75	1.68
3.9	1.07	0.27	3100	5.82	1.45
4.9	1.27	0.25	3850	6.87	1.36
5.2	1.24	0.23	4080	6.70	1.26
6.8	1.31	0.19	5340	7.08	1.02
8.4	1.33	0.15	6600	7.21	0.85
11.8	1.46	0.12	9300	7.91	0.67
16.3	1.49	0.09	12800	8.05	0.52
16.9	1.51	0.09	13300	8.16	0.51
22.6	1.59	0.07	17800	8.63	0.43
27.2	1.66	0.06	21400	8.96	0.39
33.4	1.69	0.05	26300	9.13	0.35
37.8	1.78	0.05	29700	9.61	0.34
42.9	1.80	0.05	33800	9.75	0.33

[a]Uncertainty values are for $U_h \approx B_h$ and are at a 95% confidence level.

cluded that assuming an uncertainty of $\pm15\%$ about the nominal curve seemed appropriate. This band is shown around the correlation curve in Figure 7.1.

The results for the cylinder with $L/D = 48$ are also shown in Figure 7.2 using the appropriate nondimensional variables and the coordinate system (log–log) that rectifies the data. Several points can be made concerning the data presented in Figures 7.1 and 7.2. First, the comparison at the Nth-order replication level between the experimental data and the classic results for $L/D \to \infty$ is excellent. The data and uncertainty intervals from our experiment fall within the uncertainty band around the correlation. As anticipated, the long cylinder ($L/D = 48$) approximates an infinite cylinder and provides the opportunity for a check against "known" values.

Where possible a comparison of this type should be made for any experiment. By comparing the experimental results with accepted data, or an analytical result in the limit, or an accepted correlation, the validity of the experimental apparatus and techniques can be demonstrated. In this case, since the results for $L/D = 48$ compare favorably with the classic results, the results for other L/D values obtained with the same instrumentation system and test procedure have a greater degree of credibility.

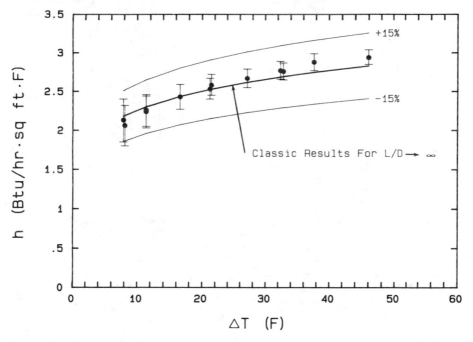

Figure 7.1 Free convective heat transfer data for a horizontal cylinder of 1/8-in. diameter and $L/D = 48$.

As shown in Figures 7.1 and 7.2, the experiment was run so that four experimental set points were replicated. The small scatter in these results relative to the size of the uncertainty intervals shows that for this experiment the precision errors are much less than the bias errors. This first-order comparison thus confirms the estimates made in the initial phases of the experiment that precision errors were negligible.

Another key point is the coordinate system used in Figure 7.2. This form of illustrating the free convection heat transfer experiment results is a case of one result, Nu, being plotted as a function of another result, GrPr, in log–log coordinates. For this case the uncertainty in the GrPr coordinate for each data point is much less than that for the Nu coordinate. For nondimensional data presentations, the quantity represented on the Y axis is generally the one with the larger uncertainty.

Nondimensional coordinates are the appropriate ones to use in most situations as they allow a broader application of the data. In this case for a similar geometric configuration ($L/D = 48$ for a horizontal cylinder) and for the same physical process (free convection heat transfer), any combination of fluid properties, diameter, and ΔT that has the same Grashof–Prandtl number will also have the same Nusselt number.

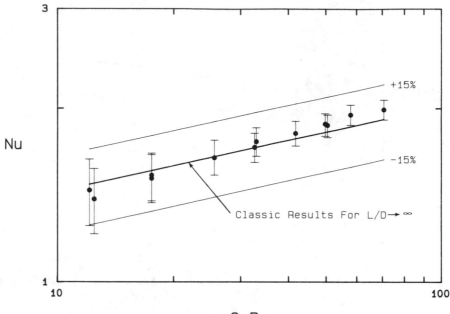

Figure 7.2 Free convective heat transfer data for a horizontal cylinder of 1/8-in. diameter and $L/D = 48$. Nondimensional rectified coordinates.

The results for the cylinder with $L/D = 6$, which are tabulated in Table 7.2, are plotted in Figure 7.3 as h versus ΔT. The curve representing the classic results for $L/D \to \infty$ is also shown on this figure with the $\pm 15\%$ uncertainty bands. We see that this short cylinder has a free convection heat transfer behavior that differs from those with large L/D.

The nondimensional presentation of the data for the $L/D = 6$ cylinder is given in Figure 7.4. Note that the uncertainty bands that are symmetrical on the linear plot (Figure 7.3) are distorted by the logarithmic scale in Figure 7.4. Also note that the small scatter in the results at the two replicated set points again indicates that precision errors are negligible with respect to bias errors in this experiment.

Since a mathematical expression to represent the $L/D = 6$ data would be useful, the functional form [Eq. (7.12)] that represents $L/D \to \infty$ was chosen for use in the regression. A linear least-squares curvefit was obtained for these data using Eqs. (7.6) and (7.7) for the slope and intercept. The Y_i and X_i quantities for these expressions are in this case log(Nu) and log(GrPr), respectively. Therefore, the curvefit expression is

$$\log(\text{Nu}) = 0.192 \log(\text{GrPr}) + 0.1232 \tag{7.13}$$

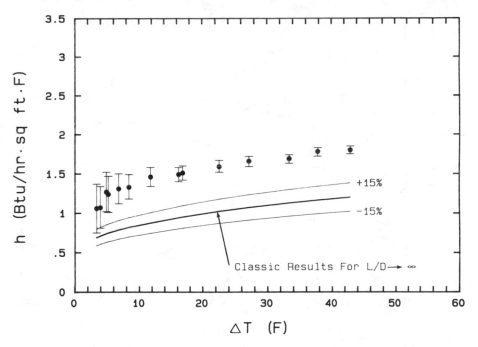

Figure 7.3 Free convective heat transfer data for a horizontal cylinder of 1-in. diameter and $L/D = 6$.

or

$$Nu = 1.328(GrPr)^{0.192} \tag{7.14}$$

The uncertainty in the Y variable in this case is not constant over the range of the curvefit, so one of the assumptions of the least-squares approach is violated. Even though the data uncertainty varies, the least-squares regression still provides a mathematical expression that well represents the data (as is immediately obvious on viewing Figure 7.4). Since our purpose in using the regression is to obtain a mathematical expression that can be used to represent the data in a compact fashion (and perhaps to interpolate between the discrete data points), the ultimate test is comparison of the regression curve with the data. If the data are well-represented, then the regression procedure can be considered successful.

The SEE was calculated for this curvefit, and the ± 2(SEE) band is plotted around the linear regression line in Figure 7.4. This band does enclose the data points, but it does not appropriately represent the uncertainty associated with using the curvefit to represent the data. As previously discussed, the correct method for determining the curvefit uncertainty is to consider the uncertainty associated with the data used in the regression. Because in this case the curvefit passes very close to all of the data points, the percentage uncertainty

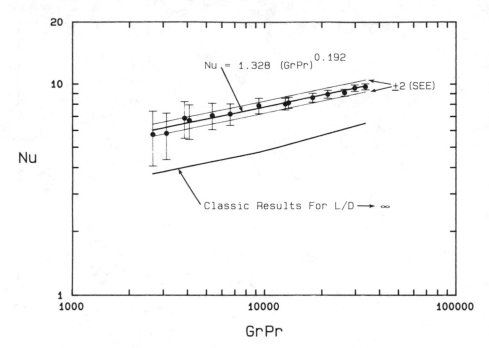

Figure 7.4 Free convective heat transfer data for a horizontal cylinder of 1-in. diameter and $L/D = 6$. Nondimensional rectified coordinates.

for each data point (calculated from the data in Table 7.2) can be used as the uncertainty associated with using the curvefit.

When reporting the results of this regression, a statement similar to the following should be made: For horizontal cylinders with $L/D = 6$ and GrPr between 2,600 and 34,000, the Nusselt number can be expressed as

$$Nu = 1.328(GrPr)^{0.192} \qquad (7.14)$$

with an uncertainty in Nu of

$$
\begin{array}{lll}
29\text{--}14\% & 2,600 & < GrPr < 5,300 \\
14\text{--}6\% & 5,300 & < GrPr < 12,000 \qquad (7.15) \\
6\text{--}3\% & 12,000 & < GrPr < 34,000
\end{array}
$$

Over each of these ranges the uncertainty can be considered linear with GrPr.

7-3.2   Summary

In this example we have discussed and illustrated some important points that should be considered in the presentation and reporting of experimental results.

The major ideas are

1. Comparisons at the first- and Nth-order replication levels should be presented if possible.
2. The results should be presented in tabular form giving the data and uncertainties (including bias and precision limits if both are significant) with the level of confidence.
3. The results should be plotted (preferably in rectified, nondimensional coordinates) with the uncertainty shown for the data points. If the uncertainty is fairly uniform for the full range of the data, then the uncertainty band can be indicated on selected data points only.
4. If a regression equation is determined for the data, then a statement of the uncertainty associated with use of the curvefit should be included. This uncertainty should be obtained by considering the uncertainty associated with the data points. Note that if the data set covers a wide range, several regression equations may be appropriate for the separate data ranges that are approximately linear. This point is illustrated in Eq. (7.12), where the correlation for cylinders with L/D approaching infinity is different for different ranges of GrPr.

7-4 MULTIPLE LINEAR REGRESSION

The basic linear least-squares analysis presented in Section 7-1 can be extended to the case in which the result of the experiment is a function of more than one variable. This dependence on more than one variable is typically the case, but we often choose to report the result as a function of one of the variables with the others held fixed. This technique was used in the previous example in which we determined a curvefit for Nu versus GrPr for a given L/D ratio. If we had more data for several other L/D ratios, we could consider obtaining a correlation equation to represent all of the results. This compact representation of the data could then be used to "predict" the free convection heat transfer coefficient for horizontal cylinders of finite length.

Consider an experimental result, R, that is a function of three variables, X, Y, and Z. If we assume that R varies linearly with respect to X, Y, and Z, we can write the expression

$$R_0 = a_1 X + a_2 Y + a_3 Z + a_4 \tag{7.16}$$

where R_0 is taken as the "optimum" R value to represent the data for a given X, Y, and Z. Just as in Section 7-1, we wish to find the values of a_1, a_2, a_3, and a_4 that minimize

$$\eta = \sum_{i=1}^{N} (R_i - a_1 X_i - a_2 Y_i - a_3 Z_i - a_4)^2 \tag{7.17}$$

where X_i, Y_i, and Z_i are the N data values.

By taking the derivatives of η with respect to the constants a_i and setting these expressions equal to zero, we obtain four equations with four unknowns

$$a_1 \sum X_i^2 + a_2 \sum X_i Y_i + a_3 \sum X_i Z_i + a_4 \sum X_i = \sum R_i X_i \tag{7.18}$$

$$a_1 \sum X_i Y_i + a_2 \sum Y_i^2 + a_3 \sum Y_i Z_i + a_4 \sum Y_i = \sum R_i Y_i \tag{7.19}$$

$$a_1 \sum X_i Z_i + a_2 \sum Y_i Z_i + a_3 \sum Z_i^2 + a_4 \sum Z_i = \sum R_i Z_i \tag{7.20}$$

$$a_1 \sum X_i + a_2 \sum Y_i + a_3 \sum Z_i + a_4 N = \sum R_i \tag{7.21}$$

where the summations run from $i = 1$ to $i = N$. In matrix notation, Eqs. (7.18)–(7.21) become

$$\begin{bmatrix} \sum X_i^2 & \sum X_i Y_i & \sum X_i Z_i & \sum X_i \\ \sum X_i Y_i & \sum Y_i^2 & \sum Y_i Z_i & \sum Y_i \\ \sum X_i Z_i & \sum Y_i Z_i & \sum Z_i^2 & \sum Z_i \\ \sum X_i & \sum Y_i & \sum Z_i & N \end{bmatrix} \begin{bmatrix} a_1 \\ a_2 \\ a_3 \\ a_4 \end{bmatrix} = \begin{bmatrix} \sum R_i X_i \\ \sum R_i Y_i \\ \sum R_i Z_i \\ \sum R_i \end{bmatrix} \tag{7.22}$$

or

$$[C][a] = [R] \tag{7.23}$$

where $[C]$, $[a]$, and $[R]$ represent the matrices in Eq. (7.22). The solution for $[a]$ can then be expressed as

$$[a] = [C]^{-1}[R] \tag{7.24}$$

where $[C]^{-1}$ is the inverse of the coefficient matrix.

The curvefit that results from this procedure is

$$R = a_1 X + a_2 Y + a_3 Z + a_4 \tag{7.25}$$

If the result is only a function of two variables (X, Y), then the terms containing Z in Eqs. (7.18)–(7.21) will be zero, and the coefficient matrix in Eq. (7.22) will be a 3×3 matrix. This procedure can also be extended for the case in which R is a function of more than three variables.

The uncertainty that should be associated with a value R calculated using the curvefit is once again related to the data points and their uncertainty. The curvefit equation should be plotted with the data points to observe how well the equation represents the data.

The condition given in Eq. (7.16) that the result R is a linear function of the independent variables may seem too restrictive at first. However, the variables X, Y, and Z do not have to be linear functions themselves, as the concept of rectification allows us to convert a nonlinear data representation equation into a linear form.

Consider the following example. Heat transfer data for external flow conditions are usually expressed as

$$\text{St} = k \, \text{Re}^a \, \text{Pr}^b \tag{7.26}$$

where St is the Stanton number, Re is the Reynolds number, Pr is the Prandtl number, and k is a constant. If we take the logarithm of both sides of Eq. (7.26) we obtain

$$\log \text{St} = \log k + a \log \text{Re} + b \log \text{Pr} \tag{7.27}$$

or in the notation and format of Eq. (7.16)

$$\log \text{St} = a_1 \log \text{Re} + a_2 \log \text{Pr} + a_3 \tag{7.28}$$

This form of representation with $\log \text{St}$ as the result and $\log \text{Re}$ and $\log \text{Pr}$ as the independent variables is in the required linear format. The curvefit constants a_i can be obtained using the technique given in Eq. (7.24).

REFERENCES

1. Schenck, H., *Theories of Engineering Experimentation*, 3rd ed., McGraw-Hill, New York, 1979.
2. *Measurement Uncertainty*, ANSI/ASME PTC 19.1-1985 Part 1, 1986.
3. Incropera, F. P., and DeWitt, D. P., *Introduction to Heat Transfer*, John Wiley, New York, 1985.

CHAPTER 7

Problems

7.1 Determine a regression equation for the following set of data. Specify the uncertainty associated with the use of this equation. Compare the ± 2

SEE band with this curvefit uncertainty.

Y	2.4	3.0	3.5	4.5	4.9	5.6	6.8	7.3
X	2.0	3.0	4.5	5.3	6.5	7.8	8.5	10.1

The uncertainty for each Y value is ± 1.

7.2 Data for the drag coefficient, C_D, for flow over an object versus Reynolds number is given below:

C_D	40	16	4.5	2	1.3	0.8	0.75	0.72	0.68	0.61
$\pm U_{C_D}$	6.0	2.2	0.60	0.24	0.14	0.08	0.07	0.06	0.04	0.03
Re	0.2	1	10	40	100	200	600	10^3	10^4	10^5

Determine the appropriate regression equation(s) for these data and specify the uncertainty associated with using the curvefit(s).

7.3 In the calibration of an orifice plate for air flow measurements, the following volumetric flow rate (Q) versus pressure drop (ΔP) data were obtained:

Q (ft^3/min)	130	304	400	488	554	610
$\pm U_Q$	11	15	16	19	17	17
ΔP (psi)	5	25	45	65	85	105

It is known that an appropriate regression equation for this situation is $Q = C(\Delta P)^{1/2}$ where Q is the flow in ft^3/min, ΔP is the orifice pressure drop in psi, and C is the calibration factor. Plot these data in rectified coordinates and determine C. What is the uncertainty in C? This uncertainty in C can be determined by first finding the uncertainty associated with Q from the curvefit over the range of ΔP values. Then over this range the uncertainty in C can be found using Eq. (3.21) with C as the result. Determine the uncertainty in C over the full range of ΔPs if the uncertainty in ΔP is 1%. Note that the uncertainty in C will vary as the uncertainty in Q varies.

APPENDIX A

USEFUL STATISTICS

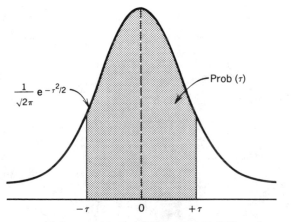

Figure A.1 Graphic representation of the two-tailed Gaussian probability.

TABLE A.1   Tabulation of Two-Tailed Gaussian Probabilities

τ	Prob(τ)	τ	Prob(τ)	τ	Prob(τ)	τ	Prob(τ)
0.00	0.0000	1.00	0.6827	2.00	0.9545	3.00	0.9973002
0.02	0.0160	1.02	0.6923	2.02	0.9566	3.05	0.9977115
0.04	0.0319	1.04	0.7017	2.04	0.9586	3.10	0.9980647
0.06	0.0478	1.06	0.7109	2.06	0.9606	3.15	0.9983672
0.08	0.0638	1.08	0.7199	2.08	0.9625	3.20	0.9986257
0.10	0.0797	1.10	0.7287	2.10	0.9643	3.25	0.9988459
0.12	0.0955	1.12	0.7373	2.12	0.9660	3.30	0.9990331
0.14	0.1113	1.14	0.7457	2.14	0.9676	3.35	0.9991918
0.16	0.1271	1.16	0.7540	2.16	0.9692	3.40	0.9993261
0.18	0.1428	1.18	0.7620	2.18	0.9707	3.45	0.9994394
0.20	0.1585	1.20	0.7699	2.20	0.9722	3.50	0.9995347
0.22	0.1741	1.22	0.7775	2.22	0.9736	3.55	0.9996147
0.24	0.1897	1.24	0.7850	2.24	0.9749	3.60	0.9996817
0.26	0.2051	1.26	0.7923	2.26	0.9762	3.65	0.9997377
0.28	0.2205	1.28	0.7995	2.28	0.9774	3.70	0.9997843
0.30	0.2358	1.30	0.8064	2.30	0.9786	3.75	0.9998231
0.32	0.2510	1.32	0.8132	2.32	0.9797	3.80	0.9998552
0.34	0.2661	1.34	0.8198	2.34	0.9807	3.85	0.9998818
0.36	0.2812	1.36	0.8262	2.36	0.9817	3.90	0.9999037
0.38	0.2961	1.38	0.8324	2.38	0.9827	3.95	0.9999218
0.40	0.3108	1.40	0.8385	2.40	0.9836	4.00	0.9999366
0.42	0.3255	1.42	0.8444	2.42	0.9845	4.05	0.9999487
0.44	0.3401	1.44	0.8501	2.44	0.9853	4.10	0.9999586
0.46	0.3545	1.46	0.8557	2.46	0.9861	4.15	0.9999667
0.48	0.3688	1.48	0.8611	2.48	0.9869	4.20	0.9999732
0.50	0.3829	1.50	0.8664	2.50	0.9876	4.25	0.9999786
0.52	0.3969	1.52	0.8715	2.52	0.9883	4.30	0.9999829
0.54	0.4108	1.54	0.8764	2.54	0.9889	4.35	0.9999863
0.56	0.4245	1.56	0.8812	2.56	0.9895	4.40	0.9999891
0.58	0.4381	1.58	0.8859	2.58	0.9901	4.45	0.9999911
0.60	0.4515	1.60	0.8904	2.60	0.9907	4.50	0.9999931
0.62	0.4647	1.62	0.8948	2.62	0.9912	4.55	0.9999946
0.64	0.4778	1.64	0.8990	2.64	0.9917	4.60	0.9999957
0.66	0.4907	1.66	0.9031	2.66	0.9922	4.65	0.9999966
0.68	0.5035	1.68	0.9070	2.68	0.9926	4.70	0.9999973
0.70	0.5161	1.70	0.9109	2.70	0.9931	4.75	0.9999979
0.72	0.5285	1.72	0.9146	2.72	0.9935	4.80	0.9999984
0.74	0.5407	1.74	0.9181	2.74	0.9939	4.85	0.9999987
0.76	0.5527	1.76	0.9216	2.76	0.9942	4.90	0.9999990
0.78	0.5646	1.78	0.9249	2.78	0.9946	4.95	0.9999992
0.80	0.5763	1.80	0.9281	2.80	0.9949	5.00	0.9999994
0.82	0.5878	1.82	0.9312	2.82	0.9952		
0.84	0.5991	1.84	0.9342	2.84	0.9955		
0.86	0.6102	1.86	0.9371	2.86	0.9958		
0.88	0.6211	1.88	0.9399	2.88	0.9960		
0.90	0.6319	1.90	0.9426	2.90	0.9963		
0.92	0.6424	1.92	0.9451	2.92	0.9965		
0.94	0.6528	1.94	0.9476	2.94	0.9967		
0.96	0.6629	1.96	0.9500	2.96	0.9969		
0.98	0.6729	1.98	0.9523	2.98	0.9971		

TABLE A.2   The t Distribution[a]

ν \ C	0.900	0.950	0.990	0.995	0.999
1	6.314	12.706	63.657	127.321	636.619
2	2.920	4.303	9.925	14.089	31.598
3	2.353	3.182	5.841	7.453	12.924
4	2.132	2.776	4.604	5.598	8.610
5	2.015	2.571	4.032	4.773	6.869
6	1.943	2.447	3.707	4.317	5.959
7	1.895	2.365	3.499	4.029	5.408
8	1.860	2.306	3.355	3.833	5.041
9	1.833	2.262	3.250	3.690	4.781
10	1.812	2.228	3.169	3.581	4.587
11	1.796	2.201	3.106	3.497	4.437
12	1.782	2.179	3.055	3.428	4.318
13	1.771	2.160	3.012	3.372	4.221
14	1.761	2.145	2.977	3.326	4.140
15	1.753	2.131	2.947	3.286	4.073
16	1.746	2.120	2.921	3.252	4.015
17	1.740	2.110	2.898	3.223	3.965
18	1.734	2.101	2.878	3.197	3.922
19	1.729	2.093	2.861	3.174	3.883
20	1.725	2.086	2.845	3.153	3.850
21	1.721	2.080	2.831	3.135	3.819
22	1.717	2.074	2.819	3.119	3.792
23	1.714	2.069	2.807	3.104	3.768
24	1.711	2.064	2.797	3.090	3.745
25	1.708	2.060	2.787	3.078	3.725
26	1.706	2.056	2.779	3.067	3.707
27	1.703	2.052	2.771	3.057	3.690
28	1.701	2.048	2.763	3.047	3.674
29	1.699	2.045	2.756	3.038	3.659
30	1.697	2.042	2.750	3.030	3.646
40	1.684	2.021	2.704	2.971	3.551
60	1.671	2.000	2.660	2.915	3.460
120	1.658	1.980	2.617	2.860	3.373
∞	1.645	1.960	2.576	2.807	3.291

[a] Given are the values of t for a confidence level C and number of degrees of freedom $\nu = N - 1$.

APPENDIX B

PROPAGATION OF ERRORS INTO AN EXPERIMENTAL RESULT

In most experimental programs, the experimental result r is not measured directly. It is generally related to measured variables X_i through a data reduction equation

$$r = r(X_1, X_2, \ldots, X_J) \qquad \text{(B.1)}$$

that is not necessarily linear. Because all of the measurements of the X_is will contain errors, we wish to determine how these errors are manifested in the result. In other words, "How do the errors in the measured variables propagate through the data reduction expression into the result?"

B-1 OVERVIEW

As Mandel [1] points out, the exact calculation of the variance of a nonlinear function of variables that are subject to error is a problem of considerable mathematical complexity. Fortunately, in practice an exact determination is rarely necessary, and approximate estimates using a linearized approach are adequate in most applications. The linearized approach is based on the use of a Taylor series expansion of the function [Eq. (B.1)] with only the linear terms retained.

In Section B-2, the derivation of such an approach that considers the propagation of the precision errors in the measured variables into the result is presented. The assumptions and approximations that are made are explicitly pointed out and discussed.

188

In Section B-3, the different interpretations [2, 3, 4, 5] of the propagation equation that have been made are examined and discussed.

In Section B-4, an approximate analysis is presented that investigates the assumptions involved in the separate propagation of the biases and the precision errors and their final combination in the result as recommended in the Standard [5].

B-2   PROPAGATION OF PRECISION ERRORS INTO AN EXPERIMENTAL RESULT

Rather than presenting the derivation for the case in which the result is a function of many variables, the simpler case in which the result is a function of only two variables will be considered first. The expressions for the more general case will then be presented as extensions of the two-variable case.

B-2.1   Result a Function of Two Variables

For this case the data reduction equation is

$$r = r(x, y) \tag{B.2}$$

We assume that the function is continuous and has continuous derivatives in the domain of interest.

Suppose that at a given "constant" experimental condition (or set point) a large number, N, of data pairs x_i, y_i are measured and for each data pair a value of the result r_i is calculated using Eq. (B.2). This situation is illustrated in Figure B.1 for a single data pair. Each measurement of x is biased by a constant amount β_x and contains a random error ϵ_{x_i}, whereas each measurement of y is biased by a constant amount β_y and contains a random error ϵ_{y_i}. The random errors in x_i and y_i cause random errors (ϵ_{r_i}) in the result r_i. Note that the means of the parent populations of x_i and y_i are μ_x and μ_y, respectively.

In Chapter 2 the use of the precision index as a descriptor or measure of the random errors in measurements of a single variable is discussed. In what we are considering here, we want to determine and quantify the effect of random errors in measured variables on the random error in the result. Another way of stating this is that we want to determine how random errors in measurements propagate into the experimental result. What is the relationship between the precision index of the result and the precision indices of the measured variables?

To investigate this question, the function in Eq. (B.2) is approximated using a Taylor series expansion. Because we are interested in the propagation of the random errors, the expansion is centered at the biased result $r(\mu_x, \mu_y)$ rather than r_{true}. Expanding to the general point (x_i, y_i) in the neighborhood of

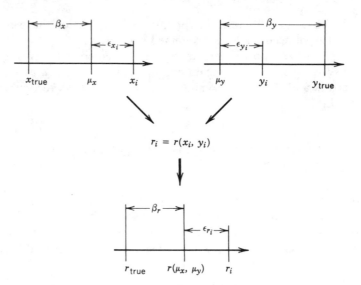

Figure B.1   Propagation of bias errors and precision errors into an experimental result.

(μ_x, μ_y) gives

$$r(x_i, y_i) = r(\mu_x, \mu_y) + \frac{\partial r}{\partial x}(x_i - \mu_x) + \frac{\partial r}{\partial y}(y_i - \mu_y) + R_2 \quad \text{(B.3)}$$

where the partial derivatives are evaluated at (μ_x, μ_y). R_2, the remainder after two terms, is [6]

$$R_2 = \frac{1}{2!}\left[\frac{\partial^2 r}{\partial x^2}(x_i - \mu_x)^2 + 2\frac{\partial^2 r}{\partial x\,\partial y}(x_i - \mu_x)(y_i - \mu_y)\right.$$

$$\left. + \frac{\partial^2 r}{\partial y^2}(y_i - \mu_y)^2\right] \quad \text{(B.4)}$$

where the partial derivatives are evaluated at (ξ, η), which is somewhere between (x_i, y_i) and (μ_x, μ_y).

In a linearized approach to error propagation, we assume that R_2 is zero (or at least negligible) so that only the first-order, or linear, terms in the expansion are retained and

$$r_i = r(x_i, y_i) = r(\mu_x, \mu_y) + \frac{\partial r}{\partial x}(x_i - \mu_x) + \frac{\partial r}{\partial y}(y_i - \mu_y) \quad \text{(B.5)}$$

If the function in Eq. (B.2) is linear

$$r = ax + by \tag{B.6}$$

then the partial derivatives in (B.4) are identically zero and (B.5) is exact. If the function is nonlinear, then we implicitly assume that R_2 is negligible when we use (B.5). This is a reasonable assumption if the $(x_i - \mu_x)$ and $(y_i - \mu_y)$ are "sufficiently" small and/or if the partial derivatives in (B.4) are small.

Rearranging Eq. (B.5) as

$$r(x_i, y_i) - r(\mu_x, \mu_y) = \frac{\partial r}{\partial x}(x_i - \mu_x) + \frac{\partial r}{\partial y}(y_i - \mu_y) \tag{B.7}$$

and noting from Figure B.1 that

$$\epsilon_{r_i} = r(x_i, y_i) - r(\mu_x, \mu_y)$$

$$\epsilon_{x_i} = x_i - \mu_x \tag{B.8}$$

$$\epsilon_{y_i} = y_i - \mu_y$$

we obtain the relationship between the random error in the result and the random errors in the measured variables as

$$\epsilon_{r_i} = \frac{\partial r}{\partial x}\epsilon_{x_i} + \frac{\partial r}{\partial y}\epsilon_{y_i} \tag{B.9}$$

It should be remembered that the partial derivatives are evaluated at (μ_x, μ_y) so that they are constant for all i.

Now we square the random errors and sum them for the N measurements to obtain

$$\sum_{i=1}^{N} (\epsilon_{r_i})^2 = \sum_{i=1}^{N} \left[\frac{\partial r}{\partial x}\epsilon_{x_i} + \frac{\partial r}{\partial y}\epsilon_{y_i} \right]^2$$

$$= \sum_{i=1}^{N} \left(\frac{\partial r}{\partial x} \right)^2 (\epsilon_{x_i})^2 + \sum_{i=1}^{N} \left(\frac{\partial r}{\partial y} \right)^2 (\epsilon_{y_i})^2$$

$$+ \sum_{i=1}^{N} 2 \left(\frac{\partial r}{\partial x} \right) \left(\frac{\partial r}{\partial y} \right) \epsilon_{x_i} \epsilon_{y_i} \tag{B.10}$$

Now divide by N, let N approach infinity, and recall the definition of the variance of a distribution as

$$\sigma_x^2 = \lim_{N \to \infty} \frac{1}{N} \sum_{i=1}^{N} (x_i - \mu_x)^2 \tag{B.11}$$

so that Eq. (B.10) can be rewritten in the form

$$\sigma_r^2 = \left(\frac{\partial r}{\partial x} \right)^2 \sigma_x^2 + \left(\frac{\partial r}{\partial y} \right)^2 \sigma_y^2 + 2 \left(\frac{\partial r}{\partial x} \right) \left(\frac{\partial r}{\partial y} \right) \sigma_{xy} \tag{B.12}$$

where σ_{xy} is the *covariance* of x and y defined as

$$\sigma_{xy} = \lim_{N \to \infty} \frac{1}{N} \sum_{i=1}^{N} (x_i - \mu_x)(y_i - \mu_y)$$

$$= \lim_{N \to \infty} \frac{1}{N} \sum_{i=1}^{N} \epsilon_{x_i} \epsilon_{y_i} \tag{B.13}$$

If the errors ϵ_{x_i} and ϵ_{y_i} are independent of each other, then the covariance is zero. If they are not independent, then the covariance can be either positive or negative. If, for instance, ϵ_{y_i} tends to be positive if ϵ_{x_i} is positive and negative if ϵ_{x_i} is negative, then the covariance would be nonzero and positive. The covariance is often normalized to form a coefficient of correlation between x and y

$$\rho_{xy} = \frac{\sigma_{xy}}{\sigma_x \sigma_y} \tag{B.14}$$

This correlation coefficient always lies between -1 and $+1$ and is zero when the ϵ_{x_i} and ϵ_{y_i} are independent of each other.

Using Eq. (B.14), we can express Eq. (B.12) in the form

$$\sigma_r^2 = \left(\frac{\partial r}{\partial x} \right)^2 \sigma_x^2 + \left(\frac{\partial r}{\partial y} \right)^2 \sigma_y^2 + 2 \left(\frac{\partial r}{\partial x} \right) \left(\frac{\partial r}{\partial y} \right) \rho_{xy} \sigma_x \sigma_y \tag{B.15}$$

If the errors in x and y are not correlated, then ρ_{xy} is zero and

$$\sigma_r^2 = \left(\frac{\partial r}{\partial x} \right)^2 \sigma_x^2 + \left(\frac{\partial r}{\partial y} \right)^2 \sigma_y^2 \tag{B.16}$$

This is the error propagation or uncertainty analysis equation as it has usually been derived or presented [2, 3, 4, 5]. There have been some differences in interpretation when the equation is applied, however, and these are discussed in Section B-3.

B-2.2   Result a Function of Many Variables

For the more general case in which the result is a function of many variables, the data reduction equation can be expressed in the form

$$r = r(X_1, X_2, \ldots, X_J)$$
(B.17)

Using the definitions

$$\theta_i = \frac{\partial r}{\partial X_i}$$
(B.18)

Eq. (B.15) can be extended to the more general case to yield

$$\sigma_r^2 = \sum_{i=1}^{J} \sum_{k=1}^{J} \theta_i \theta_k \rho_{ik} \sigma_i \sigma_k$$
(B.19)

where

$$\rho_{ik} \equiv 1.0 \qquad \text{for } i = k$$
(B.20)

This can also be written in the form

$$\sigma_r^2 = \sum_{i=1}^{J} \left[\theta_i^2 \sigma_i^2 + \sum_{k=1}^{J} \theta_i \theta_k \rho_{ik} \sigma_i \sigma_k (1 - \delta_{ik}) \right]$$
(B.21)

where the Kronecker delta is

$$\delta_{ik} = \begin{cases} 1 & i = k \\ 0 & i \neq k \end{cases}$$
(B.22)

In cases in which each of the standard deviations is independent of all of the other standard deviations, these expressions reduce to

$$\sigma_r^2 = \sum_{i=1}^{J} \theta_i^2 \sigma_i^2$$
(B.23)

which is the more general equivalent of Eq. (B.16).

B-3   DIFFERENT INTERPRETATIONS OF THE PROPAGATION EQUATION

If the precision errors in all of the variables are normally distributed, then Eq. (B.23) describes the propagation of 68% confidence intervals since $\pm \sigma$ encloses about 68% of a Gaussian population. From another viewpoint, the propagation is at approximately 2 to 1 odds. Kline and McClintock [2] extended Eq. (B.23) by replacing the standard deviations with general uncertainty intervals, w, so that

$$w_r^2 = \sum_{i=1}^{J} \theta_i^2 w_i^2 \tag{B.24}$$

or, for the two-variable case,

$$w_r^2 = \left(\frac{\partial r}{\partial x} \right)^2 w_x^2 + \left(\frac{\partial r}{\partial y} \right)^2 w_y^2 \tag{B.25}$$

They stipulated that the variables x and y be normally distributed and that w_x and w_y be intervals for the same odds (or confidence). The odds on w_r are then the same as on w_x and w_y. Stated another way, if

$$w_x = tS_x \tag{B.26}$$

where t is chosen for 95% confidence, then

$$w_y = tS_y \tag{B.27}$$

where t must also be chosen for 95% confidence and w_r from Eq. (B.25) will be a 95% confidence interval for the uncertainty in the result.

Schenck [3] uses the same interpretation, whereas Taylor [4] chooses to use the standard deviation as the measure of the uncertainty and hence stays with Eq. (B.23).

A somewhat different interpretation is outlined and recommended in the Standard [5]. Eq. (B.23) is derived in the same form as previously. However, the standard deviations are replaced by the precision indices of the measured variables so that

$$S_r^2 = \sum_{i=1}^{J} \theta_i^2 S_i^2 \tag{B.28}$$

Once the precision indices have been propagated, the number of degrees of freedom in S_r is determined from the Welch–Satterthwaite formula (see Chapter 4). The resulting number of degrees of freedom is then used with the t-distribution to give the value of t for 95% confidence, and the portion of the uncertainty in the result *due to precision error* is then calculated as (tS_r).

What are the differences between these two approaches to the propagation of precision errors? First, if all of the precision indices S_i are determined from (or estimated for) large sample sizes so that $t = 2$ in every instance, then the two approaches are exactly the same.

The differences occur when the precision indices for the measured variables are determined for different small ($N < 30$) sample sizes. In such cases, the propagation of the S_is in Eq. (B.28) is not at a constant confidence level (not at the same odds). This is "corrected for" by using the Welch–Satterthwaite formula to determine the number of degrees of freedom in S_r. This information, used with the t-distribution, provides the value of t such that $\pm(tS_r)$ is a 95% confidence interval. The procedure outlined in the Standard also specifies that the bias limits B_i, which are the 95% confidence estimates of the bias errors in the measured variables, be propagated by replacing the σ_is in Eq. (B.23) with the B_is so that

$$B_r^2 = \sum_{i=1}^{J} \theta_i^2 B_i^2 \tag{B.29}$$

The uncertainty in the result for 95% confidence is then determined by combining the bias and precision contributions using

$$U_r^2 = B_r^2 + (tS_r)^2 \tag{B.30}$$

The extension of Eq. (B.23) to propagate the bias limits is not obvious, conceptually, because the equation was derived by using a Taylor series expansion about the biased value $r(\mu_x, \mu_y)$. This extension is discussed in Section B-4.

The Standard's treatment of the propagation of bias differs from previous interpretations. Schenck [3, p. 78] assumed that fixed error is eliminated through calibration, leaving only the propagation of precision error to be considered. Taylor [4, p. 175] made a similar assumption. Kline and McClintock [2] considered the biases that remain after elimination of those that are known to be random errors and thus contained in the w_is. This view considers, for example, the bias in an instrument to be a single sample (of unknown value) from the distribution of biases in a collection of "identical" instruments. This corresponds to the Nth-order level of replication described by Moffat [7].

The U_r of the Standard approach [Eq. (B.30)] and the w_r of the other approaches [Eq. (B.24)] are identical for cases in which 95% confidence is used for all estimates, t is equal to two in all instances, and the w_is are interpreted as

$$w_i^2 = B_i^2 + (tS_i)^2 \tag{B.31}$$

B-4   PROPAGATION OF BOTH BIAS AND PRECISION ERRORS INTO AN EXPERIMENTAL RESULT

In this section, we consider the propagation of both bias and precision errors into an experimental result. The approach taken is similar to that in Section

B-2, although here the Taylor series expansion is made about the true result rather than the biased result.

Consider the situation shown in Figure B.1 again, except this time let the biases be random variables so that

$$x_i = x_{true} + b_{x_i} + \epsilon_{x_i} \tag{B.32}$$

and

$$y_i = y_{true} + b_{y_i} + \epsilon_{y_i} \tag{B.33}$$

This is the case that would occur if each measurement x_i were to be made with a different instrument from the population of "identical" instruments of the model being used to measure x and similarly for the variable y.

The function in the data reduction equation

$$r = r(x, y) \tag{B.34}$$

is approximated using a Taylor series expansion. Expanding to the general point (x_i, y_i) in the neighborhood of (x_{true}, y_{true}) gives

$$r(x_i, y_i) = r(x_{true}, y_{true}) + \frac{\partial r}{\partial x}(x_i - x_{true}) + \frac{\partial r}{\partial y}(y_i - y_{true}) \tag{B.35}$$

where the remainder term has been assumed negligible and the partial derivatives are evaluated at (x_{true}, y_{true}). Because the true values are always unknown, an additional approximation is always made in practice when the partial derivatives are evaluated at a chosen point. Rearranging Eq. (B.35) as

$$r(x_i, y_i) - r(x_{true}, y_{true}) = \frac{\partial r}{\partial x}(x_i - x_{true}) + \frac{\partial r}{\partial y}(y_i - y_{true}) \tag{B.36}$$

and using the definitions

$$\delta_{r_i} = r(x_i, y_i) - r(x_{true}, y_{true})$$

$$b_{x_i} + \epsilon_{x_i} = x_i - x_{true}$$

$$b_{y_i} + \epsilon_{y_i} = y_i - y_{true} \tag{B.37}$$

$$\theta_x = \partial r / \partial x$$

$$\theta_y = \partial r / \partial y$$

we obtain the relationship between the error in the result and the bias and precision errors in the measured variables as

$$\delta_{r_i} = \theta_x b_{x_i} + \theta_x \epsilon_{x_i} + \theta_y b_{y_i} + \theta_y \epsilon_{y_i} \tag{B.38}$$

Now we square the errors in the result and sum them for the N measurements to obtain

$$\sum_{i=1}^{N} (\delta_{r_i})^2 = \theta_x^2 \sum_{i=1}^{N} b_{x_i}^2 + \theta_y^2 \sum_{i=1}^{N} b_{y_i}^2 + \theta_x^2 \sum_{i=1}^{N} \epsilon_{x_i}^2 + \theta_y^2 \sum_{i=1}^{N} \epsilon_{y_i}^2$$

$$+ 2\theta_x^2 \sum_{i=1}^{N} b_{x_i} \epsilon_{x_i} + 2\theta_x \theta_y \sum_{i=1}^{N} b_{x_i} b_{y_i} + 2\theta_x \theta_y \sum_{i=1}^{N} b_{x_i} \epsilon_{y_i}$$

$$+ 2\theta_y^2 \sum_{i=1}^{N} b_{y_i} \epsilon_{y_i} + 2\theta_x \theta_y \sum_{i=1}^{N} b_{y_i} \epsilon_{x_i}$$

$$+ 2\theta_x \theta_y \sum_{i=1}^{N} \epsilon_{x_i} \epsilon_{y_i} \tag{B.39}$$

If it is assumed that the precision errors (ϵs) and bias errors (bs) are always uncorrelated, the four terms with ($\Sigma\, b\epsilon$) are zero for large N and

$$\sum_{i=1}^{N} (\delta_{r_i})^2 = \theta_x^2 \sum_{i=1}^{N} b_{x_i}^2 + \theta_y^2 \sum_{i=1}^{N} b_{y_i}^2 + 2\theta_x \theta_y \sum_{i=1}^{N} b_{x_i} b_{y_i}$$

$$+ \theta_x^2 \sum_{i=1}^{N} \epsilon_{x_i}^2 + \theta_y^2 \sum_{i=1}^{N} \epsilon_{y_i}^2 + 2\theta_x \theta_y \sum_{i=1}^{N} \epsilon_{x_i} \epsilon_{y_i} \tag{B.40}$$

Now define

$$U_r = \left[\frac{1}{N} \sum_{i=1}^{N} (\delta_{r_i})^2 \right]^{1/2} \tag{B.41}$$

as the uncertainty in the result *at some confidence level that is unknown at this point.* Then

$$U_r^2 = \frac{1}{N} \sum_{i=1}^{N} (\delta_{r_i})^2 = \theta_x^2 \frac{1}{N} \sum_{i=1}^{N} b_{x_i}^2 + \theta_y^2 \frac{1}{N} \sum_{i=1}^{N} b_{y_i}^2$$

$$+ 2\theta_x \theta_y \frac{1}{N} \sum_{i=1}^{N} b_{x_i} b_{y_i} + \theta_x^2 \frac{1}{N} \sum_{i=1}^{N} \epsilon_{x_i}^2$$

$$+ \theta_y^2 \frac{1}{N} \sum_{i=1}^{N} \epsilon_{y_i}^2 + 2\theta_x \theta_y \frac{1}{N} \sum_{i=1}^{N} \epsilon_{x_i} \epsilon_{y_i} \tag{B.42}$$

Letting N approach infinity, Eq. (B.42) can be written

$$U_r^2 = \theta_x^2 B_x^2 + \theta_y^2 B_y^2 + 2\theta_x\theta_y B_{xy}$$
$$+\theta_x^2\sigma_x^2 + \theta_y^2\sigma_y^2 + 2\theta_x\theta_y\sigma_{xy} \tag{B.43}$$

where (B_x^2) and (B_y^2) are the variances of the bias error distributions, (B_{xy}) is the covariance of the bias error distributions, and the σs are the variances and covariance of the precision error distributions as previously defined in Eqs. (B.11) and (B.13).

If additional definitions are made such that

$$B_r^2 = \theta_x^2 B_x^2 + \theta_y^2 B_y^2 + 2\theta_x\theta_y B_{xy} \tag{B.44}$$

and

$$\sigma_r^2 = \theta_x^2\sigma_x^2 + \theta_y^2\sigma_y^2 + 2\theta_x\theta_y\sigma_{xy} \tag{B.45}$$

then

$$U_r^2 = B_r^2 + \sigma_r^2 \tag{B.46}$$

At this point, let us assume that all of the error distributions are Gaussian and also recall that there is no way to measure the bias errors (all bias errors whose values can be determined are assumed to have been eliminated by calibration corrections, etc.). Then σ_x and σ_y are the 68% confidence intervals for the precision errors and it makes sense to interpret (B_x) and (B_y) as 68% confidence estimates for the bias errors. This means that the uncertainty in the result in Eq. (B.46) should be considered as a 68% coverage estimate.

Extending this concept in the same manner as Kline and McClintock [2], a 95% coverage estimate of the uncertainty in the result can be expressed as

$$U_r^2 = B_r^2 + (tS_r)^2 \tag{B.47}$$

Here

$$B_r^2 = \theta_x^2 B_x^2 + \theta_y^2 B_y^2 + 2\theta_x\theta_y B_x B_y \rho_{xy} \tag{B.48}$$

where B_x and B_y are the 95% confidence estimates of the bias limits and ρ_{xy} is the correlation coefficient appropriate for the bias errors, and

$$(tS_r)^2 = \theta_x^2(tS_x)^2 + \theta_y^2(tS_y)^2 + 2\theta_x\theta_y(tS_x)(tS_y)\rho_{xy} \tag{B.49}$$

where (tS_x) and (tS_y) are the 95% confidence intervals for the precision errors (i.e., the precision limits P_x and P_y) and ρ_{xy} is the correlation coefficient appropriate for the precision errors.

Extension of Eqs. (B.48) and (B.49) to the case in which the result is a function of many variables yields

$$B_r^2 = \sum_{i=1}^{J} \left[\theta_i^2 B_i^2 + \sum_{k=1}^{J} \theta_i \theta_k \rho_{ik} B_i B_k (1 - \delta_{ik}) \right]$$

and

$$(tS_r)^2 = \sum_{i=1}^{J} \left[\theta_i^2 (tS_i)^2 + \sum_{k=1}^{J} \theta_i \theta_k \rho_{ik} (tS_i)(tS_k)(1 - \delta_{ik}) \right]$$

where the Kronecker delta is as defined in Eq. (B.22).

These expressions are equivalent to those recommended in the Standard for the case in which all of the sample sizes are large so that all of the ts are equal to two. The approach recommended in the Standard for the propagation of the precision indices for cases in which all of the ts are not equal is discussed in Section B-3 and in Chapter 4.

REFERENCES

1. Mandel, J., *The Statistical Analysis of Experimental Data*, Interscience–John Wiley, New York, 1964.
2. Kline, S. J., and McClintock, F. A., "Describing Uncertainties in Single-Sample Experiments," *Mechanical Engineering*, Vol 75, Jan. 1953, pp 3–8.
3. Schenck, H., *Theories of Engineering Experimentation*, 3rd ed., McGraw-Hill, New York, 1979.
4. Taylor, J. R., *An Introduction to Error Analysis: The Study of Uncertainties in Physical Measurements*, University Science Books, Mill Valley, CA, 1982.
5. *Measurement Uncertainty*, ANSI/ASME PTC 19.1-1985 Part 1. (Available from ASME Order Dept., 22 Law Drive, Box 2300, Fairfield, New Jersey 07007-2300.)
6. Hildebrand, F. B., *Advanced Calculus for Applications*, Prentice-Hall, Englewood Cliffs, NJ, 1962.
7. Moffat, R. J., "Contributions to the Theory of Single-Sample Uncertainty Analysis," *J. Fluids Engineering*, Vol. 104, June 1982, pp 250–260.

INDEX